Nikola Staykov

The Role of GABP-a/b In the Proliferation of NIH-3T3 Cells

Nikola Staykov

The Role of GABP-a/b In the Proliferation of NIH-3T3 Cells

One search of an expected function, leading to rather unexpected finds

Südwestdeutscher Verlag für Hochschulschriften

Impressum/Imprint (nur für Deutschland/only for Germany)
Bibliografische Information der Deutschen Nationalbibliothek: Die Deutsche Nationalbibliothek verzeichnet diese Publikation in der Deutschen Nationalbibliografie; detaillierte bibliografische Daten sind im Internet über http://dnb.d-nb.de abrufbar.

Alle in diesem Buch genannten Marken und Produktnamen unterliegen warenzeichen-, marken- oder patentrechtlichem Schutz bzw. sind Warenzeichen oder eingetragene Warenzeichen der jeweiligen Inhaber. Die Wiedergabe von Marken, Produktnamen, Gebrauchsnamen, Handelsnamen, Warenbezeichnungen u.s.w. in diesem Werk berechtigt auch ohne besondere Kennzeichnung nicht zu der Annahme, dass solche Namen im Sinne der Warenzeichen- und Markenschutzgesetzgebung als frei zu betrachten wären und daher von jedermann benutzt werden dürften.

Coverbild: www.ingimage.com

Verlag: Südwestdeutscher Verlag für Hochschulschriften GmbH & Co. KG
Heinrich-Böcking-Str. 6-8, 66121 Saarbrücken, Deutschland
Telefon +49 681 37 20 271-1, Telefax +49 681 37 20 271-0
Email: info@svh-verlag.de

Approved by: Würzburg, Julius-Maximilians-Universität, Diss., 2012

Herstellung in Deutschland:
Schaltungsdienst Lange o.H.G., Berlin
Books on Demand GmbH, Norderstedt
Reha GmbH, Saarbrücken
Amazon Distribution GmbH, Leipzig
ISBN: 978-3-8381-3198-6

Imprint (only for USA, GB)
Bibliographic information published by the Deutsche Nationalbibliothek: The Deutsche Nationalbibliothek lists this publication in the Deutsche Nationalbibliografie; detailed bibliographic data are available in the Internet at http://dnb.d-nb.de.

Any brand names and product names mentioned in this book are subject to trademark, brand or patent protection and are trademarks or registered trademarks of their respective holders. The use of brand names, product names, common names, trade names, product descriptions etc. even without a particular marking in this works is in no way to be construed to mean that such names may be regarded as unrestricted in respect of trademark and brand protection legislation and could thus be used by anyone.

Cover image: www.ingimage.com

Publisher: Südwestdeutscher Verlag für Hochschulschriften GmbH & Co. KG
Heinrich-Böcking-Str. 6-8, 66121 Saarbrücken, Germany
Phone +49 681 37 20 271-1, Fax +49 681 37 20 271-0
Email: info@svh-verlag.de

Printed in the U.S.A.
Printed in the U.K. by (see last page)
ISBN: 978-3-8381-3198-6

Copyright © 2012 by the author and Südwestdeutscher Verlag für Hochschulschriften GmbH & Co. KG and licensors
All rights reserved. Saarbrücken 2012

TABLE OF CONTENTS

1. INTRODUCTION 1
   1.1 The *Ets* Factor Family 1
   1.2 GABP as an *Ets* Factor 3
      1.2.1 GA-Binding Protein – Complex of GABP• and GABP• 3
      1.2.2 The GABP• Subunit 7
      1.2.3 The GABP• Subunit 10
   1.3 GABP• •• Target Genes – from *in vitro* to *in vivo* 13
   1.4 GABP in Signalling Processes 20
   1.5 GABP Protein – Protein Interactions – Partner Proteins 21
      1.5.1 Interactions with GABP• 22
         1.5.1.1. Sp1 22
         1.5.1.2. PU.1 23
         1.5.1.3. ATF1 23
         1.5.1.4. p300/CBP 24
         1.5.1.5. MITF 24
      1.5.2 Interactions with GABP• 25
         1.5.2.1. HCF-1 25
         1.5.2.2. YEAF1/YAF-2 25
         1.5.2.3. E2F1 26
   1.6 The Aim and Experimental Design of the Project 27

2. RESULTS 28
   2.1 The *Gabp•* Gene Expression Effectively Silenced by RNAi Techniques ... 28
   2.2 Constitutive Overexpression of Exogenous GABP• Accelerates Proliferation of NIH-3T3 Cells 30
      2.2.1 Transgenic NIH-3T3 Cells Are Stably Overexpressing Exogenous GABP• 30
      2.2.2 The Overexpressed GABP• Protein Demonstrates Effective Nuclear Targeting 31
      2.2.3 Constitutive GABP• Overexpression Accelerates Cell Proliferation ... 32

2.3 Newly Created Vectors Direct Simultaneous Doxycycline-Regulated Expression of GABP• and GABP• 33
2.4 Stably Transfected NIH-3T3 Cells Conditionally Express High Amounts of GABP• and GABP• 35
   2.4.1 Transgenic NIH-3T3-S2 Cells Constitutively Express S2 Transactivator 36
   2.4.2 Double Transfected Cells Overexpress GABP• •• 38
2.5 The Inducible GABP• •• Proteins Demonstrate High Levels of Overexpression and Effective Nuclear Targeting 39
2.6 The Exogenous GABP• •• Subunits Possess Strong DNA-Binding Activity *in Vitro* 40
2.7 Expression of Excessive Amounts GABP• •• Influences the Proliferation Speed of NIH-3T3 Cells 41
   2.7.1 Conditionally Increasing the GABP• •• Expression Does Not Influence the Size of NIH-3T3 Cells 41
   2.7.2 Conditionally Elevated GABP• •• Expression Results in a Reversible Reduction of Proliferation Speed of NIH-3T3 Cells 42
2.8 GABP• •• Overexpression Does Not Affect the Cell Cycle 48
2.9 Excessive Expression of GABP• •• Increases Apoptotic Processes in NIH-3T3 Cell Cultures 50
   2.9.1 Increased Sub-G_1 Cell Cycle Population Indicates Increased Apoptosis in GABP• •• Overexpressing Cell Cultures 50
   2.9.2 Examinations of the Caspase Pathways Reveal Elevated Expression and Activation of Some of Their Key Members 51
   2.9.3. Increased Expression of GABP• •• *per se* does Not Induce ER-Stress 52

3. DISCUSSION 55

3.1 Efficient Downregulation of GABP• Expression by RNAi Could Be Achieved Only Transiently 55

3.2 Constitutive GABP• Overexpression Speeds Up Cell Proliferation 58

3.3 Conditional Expression of Excessive GABP• /• in NIH-3T3 Cells Reduces Cell Proliferation Speed 60

3.4 Elevated expression of GABP• •• Induces Caspase-12 – Elicited Apoptosis 62

4. SUMMARY 68
ZUSAMMENFASSUNG 69

5. MATERIALS AND METHODS 70

5.1 Materials 70

    5.1.1 Instruments 70

    5.1.2 General materials 71

    5.1.3 Chemical reagents 72

    5.1.4 Solutions and Buffers 74

    5.1.5 Antibiotics 82

    5.1.6 Kits 83

    5.1.7 DNA size markers 83

    5.1.8 Protein standards 83

    5.1.9 Enzymes 83

    5.1.10 Antibodies 84

        5.1.10.1 Primary 84

        5.1.10.2 Secondary 84

    5.1.11. Oligonucleotides and primers 84

        5.1.11.1 Oligonucleotides for the construction of siRNA-expressing primary vectors 84

        5.1.11.2 Olygonucleotide for EMSA 85

        5.1.11.3 PCR primers for the amplification of inserts, used for DNA cloning 85

    5.1.12 Plasmid constructs 85

    5.1.13 Growth Media 86

        5.1.13.1 Mammalian cell culture medium for adherent cell lines 86

5.1.13.2	Bacterial culture media	86
5.1.14	Mammalian cell lines	87
5.1.15	Bacterial strains	87
5.1.16	Software	87
5.2	Methodology	87
5.2.1	DNA Methods	87
5.2.1.1	Isolation of plasmid DNA	87
5.2.1.1.1	Small scale plasmid DNA purification	87
5.2.1.1.2	Medium scale plasmid DNA purification	88
5.2.1.2	Determination of DNA/RNA concentration	88
5.2.1.3	DNA electrophoresis on agarose gel	88
5.2.1.4	Isolation of DNA from agarose gel	88
5.2.1.5	Restriction enzyme digestions of DNA	88
5.2.1.6	Polymerase chain reaction (PCR)	89
5.2.1.7	Purification of DNA	89
5.2.1.8	Ligation of DNA fragments	89
5.2.1.9	Colony hybridization	90
5.2.1.10	PI staining and Flow Cytometry	90
5.2.2	RNA Methods	91
5.2.2.1	RNA isolation from mammalian cells	91
5.2.2.2	Ribonuclease protection assay	91
5.2.3	Protein Methods	91
5.2.3.1	Preparation of protein extracts	91
5.2.3.1.1	Preparation of whole cell protein extracts	91
5.2.3.1.2	Preparation of nuclear and cytoplasmatic protein extracts	91
5.2.3.2	Determination of Protein Concentration	92
5.2.3.3	SDS-PAGE and immunodetection	92
5.2.3.3.1	SDS-polyacrylamide gel preparation and electrophoresis	92
5.2.3.3.2	Western blotting and hybridization	93
5.2.3.3.3	Stripping of nitrocellulose membrane	94
5.2.3.3.4	Cytostaining of adherent cells (NIH-3T3)	94
5.2.3.4	Luciferase assay	95
5.2.4	DNA/Protein Interaction Assays	95
5.2.4.1	Radioactive labelling and purification of DNA probe	95
5.2.4.2	Electrophoretic Mobility Shift Assay (EMSA)	96

5.2.5	Mammalian Cell Cultures	97
5.2.5.1	Maintenance of cell lines	97
5.2.5.2	Splitting of cell cultures – Adherent cells	97
5.2.5.3	Cell counting with haemocytometer	97
5.2.5.4	Crystal violet assay	97
5.2.5.5	Cell Transfections	98
5.2.5.5.1	Calcium Phosphate Transfection Method	98
5.2.5.5.2	Dendrimer-based transfection	99
5.2.5.6	Selection of transfected cells	99
5.2.5.7	Single cell cloning	99
5.2.5.8	Cryopreservation of mammalian cells	100
5.2.5.9	Thawing mammalian cells out of frozen stock	100
5.2.6	Bacterial Manipulation	100
5.2.6.1	Cultivation of bacteria (*E. coli*)	100
5.2.6.2	Preparation of E. coli competent bacterial cells	100
5.2.6.3	Transformation of E.coli competent bacterial cells	101
5.2.6.4	Cryopreservation of bacterial cells	101
5.2.7	Statistical Analyses	101

6. REFERENCES 102

7. APPENDIX 111
7.1. Abbreviation Index 111
7.2. Design of RNAi Targets 112
7.3. Clone Charts 114

1. INTRODUCTION

1.1. The *Ets* Factor Family

The *Ets* family of transcription factors is specified by the presence of a highly conservative DNA binding domain (the Ets-domain) and consists of more than 30 individual members which are closely involved in vital cellular functions, namely development, proliferation and differentiation, apoptosis and carcinogenesis (Sharrocks et al., 1997). The prototypic member of the family, Ets-1, was first identified in studies of the E26 avian erythroblastosis virus, leading to acute leukemia in chicken (Bister et al., 1982; Nunn et al. 1983; Nunn et al. 1984). *Ets* genes have been identified in many metazoan species, such as *Xenopus laevis* (Marchioni et al., 1993), *Drosophila melanogaster* (Laudet et al., 1999), *Caenorhabditis elegans* (Hart et al., 2000) and others, but not in protozoa, fungi or plants (Graves and Petersen, 1998). The number of Ets-family genes increased during evolution, with only one in the earthworm (*Lin1*), in comparison to more than 30 in mammals (Laudet et al., 1999), suggesting gene amplification and divergence of function. The structure of Ets-type DNA binding domain (DBD) is specified by an evolutionary conserved 85 amino acid long polypeptide, which adopts a "winged-helix-turn-helix" conformation, composed of three • helices and a four-stranded • sheet. This DBD binds with preference to sequences rich on purine residues and containing an obligatory GGA/T core (Ditmer et al., 1998; Ghosh et al., 1998; Bassuk et al., 1997; Wasylyk B. et al., 1998 and Graves and Petersen, 1998). In molecular dynamics simulations for the prototypic family member Ets-1, including ETS domain–DNA complexes, it was found that the specificity of the GGAA core sequence is determined by a direct readout mechanism in the protein–DNA complex (Obika et al., 2003). The major contacts occur between the helix-3 domain of Ets-1 and the oppositely lying major groove of the core DNA sequence. Additional analyses revealed that the arginine residues 391 and 394 are crucial for the binding to the GGAA sequence. In consistency with their importance these residues are found to be highly conserved in evolution. Their mode of action is characterized by creating "bidentate" connections with the GG dinucleotides from the GGAA sequence. Specificity of binding is determined by the Tyr395 residue, which creates hydrogen bonds between its hydroxyl group and specific nitrogen in the GGAA core, supporting or collapsing the bidentate contacts.

The DNA-binding capacity of Ets-1 falls under negative regulation, performed by special inhibitory areas surrounding the ETS domain from both N- and C-termini. These

Introduction

inhibitory regions in solution compress against the ETS domain and together with it create an autoinhibitory module. Its N terminus bears two α-helices, one of which unfolds only when Ets-1 binds to DNA (Garvie et al., 2002).

Different tissues express characteristic pattern of lineage-restricted and ubiquitously expressed Ets-family members. Usually, several Ets factors of both types are simultaneously expressed in a given cell type. As the DNA sequences to which they bind are very similar, it is often obscure how every individual *Ets* factor attains a specific function. However, several different specificity levels of Ets control have been identified. At the level of DNA-binding, the specificity depends on the nucleotides, flanking the central binding motif (Shore and Sharrocks, 1995). Binding of some Ets factors to low affinity binding sites might be stabilized by interaction(s) with other DNA-binding factors (Garvie et al., 2001). *Ets* transcription factors can be modified as downstream targets of several signal transduction pathways and can generate protein–protein interactions with various other transcription factors or co-activators. Employing alternative co-activators and signalling pathways are further possibilities to control specificity (Galang et al, 2004). Some Ets family members function as ternary factors – they can be recruited by other DNA binding factors (Fitzsimmons et al., 1996). Combinations of such mechanisms seem to define tissue- or developmental stage-distinctive functions for the individual *Ets* factors, resulting in a great specificity *in vivo*, in contrast to what could be detected in a different types of DNA binding assays *in vitro*. This is supported by the distinct and highly specific phenotypes of knockout mice for distinct Ets family members (Bartel et al., 2000).

The great importance of *Ets* factors for human health is exemplified by their involvement in the initiation and progression of various diseases (Ditmer et al., 1998) and malignancies. Abnormalities in the regulation of ETS-domain protein activity may play a role in the development of Down syndrome and of leukemias (Papas et al., 1990), in tumorigenesis (Sharrocks et al., 1997; Bieche et al., 2004) and tumor invasion (Trojanowska, 2000). Some evidence suggests that they may also play a role in the regulation of programmed cell death and the pathology of autoimmune diseases (Zhang et al., 1995).

1.2. GABP as an *Ets* Factor

1.2.1. *GA-Binding Protein – Complex of GABP• and GABP•*

As the name suggests GABP (GA-binding protein) factor binds to DNA sequences rich in guanine and adenine nucleotides. This is a common characteristic of the Ets core binding sequence (Thompson et al., 1991 and Sharrocks et al. 1997). As the only obligate multimeric factor, GABP is unparalleled among other Ets factors. Initially identified in studies on viral gene transcription, it is now established that GABP factors regulate genes that control cell cycle (Tanaka et al., 2002 and Imaki et al., 2003), apoptosis, differentiation and other basic and critical cellular functions. It has been reported that GABP can act as a transcriptional repressor of the ribosomal *S16* gene (Genuario and Perry, 1996), mitochondrial glycerol phosphate dehydrogenase (Hasan et al., 2002) and VEGF (Jeong et al., 2006). Furthermore, several investigations have elucidated the dual role of GABP in the expression of the *BRCA1* gene (Atlas et al., 2000). It is shown that GABP is a critical activator of BRCA1 expression by interaction with the so-called RIBS stimulatory component of the promoter. Later, a negative transcriptional element within the proximal promoter of the same gene was characterized, (UP site) which also binds GABP (Macdonald et al., 2007). In addition, the statement that GABP• is crucial for the functioning of both sites, RIBS and UP, was supported by 'knockdown' assays using a shRNA vector.

Thompson and co-workers (Thompson et al., 1991) isolated GABP at first as a rat liver specific transcription factor. Simultaneously, Watanabe and colleagues (Watanabe et al., 1988) described GABP under the name E4TF-1 (adenovirus E4 transcription factor 1) as a regulator of viral gene expression and Gugneja et al. (1995) and Virbasius et al. (1993) characterized and named it as nuclear respiratory factor (NRF-2), a mitochondrial protein, encoded in the cell's nucleus, instead in the mitochondrial genome. Other researchers have also described GABP as EF-1A (Bolwig et al., 1992), RBF-1 (Savoysky et al., 1994), XrpF1 (Marchioni et al., 1993) and • factor (Yoganathan et al., 1992). Table 1.1 lists some of the various names for GABP components, given by the different groups who published them.

Table 1.1: **GABP nomenclature, suggested by** Rosmarin et al., 2004 **and summary of the terminology to this moment**

Proposed by Rosmarin et al.	Thompson et al.	De la Brousse et al.	Watanabe et al.	Sawa et al.	Gugneja et al.
GABP•	GABP•		E4TF1-60	GABP•	NRF-2•
GABP•$_1$-42				GABP•$_2$	NRF-2•$_1$
GABP•$_1$-41	GABP•$_1$	GABP•$_1$-1	E4TF1-53	GABP•$_1$	NRF-2•$_2$
GABP•$_1$-38				GABP•$_2$	NRF-2•$_1$
GABP•$_1$-37	GABP•$_2$	GABP•$_1$-2	E4TF1-47	GABP•$_1$	NRF-2•$_2$
GABP•$_2$		GABP•$_2$-1			

Both GABP subunits are ubiquitously expressed at very high levels in liver, muscle, hematopoietic cells and brain (LaMarco et al., 1991, Brown et al., 1992 and de la Brousse et al., 1994). The concurrent expression of both subunits GABP• and GABP• throughout the whole mouse embryogenesis indicates the importance of ••• complex (O'Leary et al., 2005). In addition, alternative transcripts encoding multiple isoforms of both GABP• and •$_1$ have been identified and might provide an additional way for tissue-specific regulation of GABP transcriptional activity.

GABP is an obligate multimeric protein complex (Sharrocks et al., 1997; Sharrocks, 2001 and Oikawa et al., 2003). It consists of two distinct and unrelated proteins, GABP• and GABP•, which together form an ••• heterodimer that recognizes a DNA binding site with the 5'-GGAA-3' core. Two ••• dimers can homodimerize via the carboxy termini of the GABP• subunits and form a heterotetrameric (•••$_2$ complex, recognizing two 5'-GGAA-3' cores (Graves, 1998; Batchelor et al., 1998). The molecular weight of mouse GABP• and GABP• is 51 kDa and 41 kDa, respectively. Bachelor and colleagues (1998) suggested that the DNA binding of GABP is negatively regulated by inhibitory sequences, similarly to the auto-inhibitory loop in Ets-1 (Fig. 1.1). These authors speculated that helix 5 of GABP• is analogous in its function to helix 4 of the transcription factor Ets-1, which is also situated at the C-terminal region of the ETS domain (Donaldson et al., 1996). In Ets-1, DNA binding is negatively regulated by helix 4 (Hagman and Grosschedl, 1992; Lim et al., 1992; Wasylyk C. et al., 1992; Jonsen et al., 1996). When this protein is not bound to DNA, helix 4 is pressed against helix 1 and against two other helices, located amino-terminally to the ETS domain. The so-formed characteristic packing of these helices plays an inhibitory role and must be disrupted during DNA binding (Petersen et al., 1995; Skalicky et al., 1996). However, in GABP• the analogous helix 5 has a different mechanism of action. To hinder interaction with DNA, it does not have to bind to helix 1. Barbara J. Graves (1998) creates an activity model predicting that helix 5 inhibits the DNA binding of GABP• alone, because of its alternative position in the absence of GABP•. When helix 5 binds to GABP•, its inhibitory interactions with the *Ets* domain are prevented. Thus, without contacting GABP•, helix 5 blocks DNA binding. Obviously, both subunits are necessary for the assembly of a functional complex, which is emphasized by the presence of crucially important structures in each one of them, the DNA-binding domain (DBD) in GABP• and the transcriptional activation domain (TAD) in GABP• (LaMarco et al., 1991; Virbasius et al., 1993 and Sawada et al., 1994) (Fig. 1.1 and 1.2).

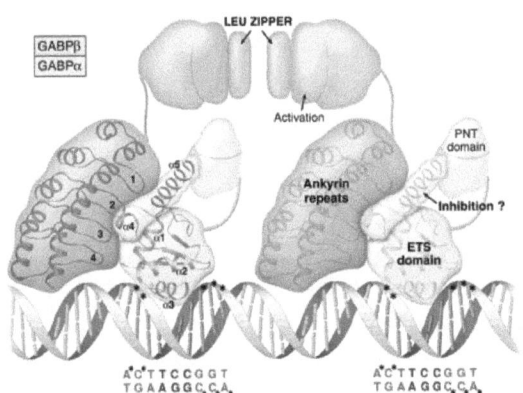

Fig. 1.1: **Drawing of the GABP heterotetramer structure with some of the functions indicated (see the text for details)**
The phosphate contacts observed in the crystal structure are denoted by asterisks. (B. Graves, Inner workings of a transcription factor partnership, Science. 1998 Feb 13;279(5353):1037-41, illustration: K. Sutliff)

GABP• binds to GABP• via four ankyrin repeats (AR), situated at its N-terminus. Each individual AR is built of two alpha helices with an intersecting loop bearing a beta turn at its tip (Batchelor et al., 1998). The tips interact each with a different part of GABP•. This includes parts of the *Ets* domain (regions from the helices 1 and 2) and the C-terminal flanking area (helices 4 and 5), part of which is involved in the inhibition of DNA-binding (helix 5). This variety of structural intra- and inter-molecular bonds illustrates the complex relations between the two interacting proteins and displays the multiple contacts mediating stable and specific connections (Graves, 1998; Batchelor et al., 1998). Specifically for the GABP complex, the separate positioning of the TAD and DBD on different subunits makes this configuration dissimilar from all *Ets* factors and rare among the other transcription factors.

GABP• binds to DNA alone with low affinity but around 100 times stronger when in complex with GABP• (LaMarco et al., 1991). Although GABP• does not directly contact DNA, it strengthens the connection by influencing the GABP• binding abilities (Graves, 1998). X-ray crystallography of bound GABP complex suggests two possible reasons why

GABP• increases the DNA binding stability (Batchelor et al., 1998). The first one is a single hydrogen bond, formed between the third ankryin repeat of GABP• and helix 1 of GABP•, which creates a contact with a phosphate from the DNA. The second reason is an interaction between helix 5 of the GABP• *Ets* domain and the first ankyrin repeat of GABP•, which blocks an inhibitory influence with the *Ets* domain (as described above). As a result, the affinity of GABP• to its corresponding binding site is strongly enhanced by the physical interaction between GABP• and GABP• and, hence, binding to the DNA of the entire complex is increased. In heterotetrameric (••$•_2$ configuration, the GABP complex identifies two 5'-GGAA-3' recognition sites and increases its DNA binding affinity even further (Thompson et al., 1991; Virbasius et al., 1993; de la Brousse et al., 1994) (Fig. 1.1). The binding can be inhibited by DNA methylation within the GABP binding site under certain conditions (Thompson et al., 1991 and Lucas et al., 2009).

GABP• and • subunits are localized in both the nucleus and cytoplasm (Yang et al., 2004). In the nucleus they are known to be associated primarily with euchromatin rather than heterochromatin, consistent with an active involvement in transcription. In cytoplasm, they are associated mainly with free ribosomes and occasionally with the Golgi apparatus and the outer membrane of the nuclear envelope. None of the subunits was found in the mitochondria (Yang et al., 2004). In primary neuronal cultures, the •-subunit of GABP was localized immunocytochemically in both the cytoplasm and the nucleus, whereas the •-subunit was present mainly in the nucleus. It was found that neuronal activity regulates subunit concentrations of GABP in the nucleus (Zhang and Wong-Riley, 2000). In cultured rat visual cortical neurons upon activation a rapid, six- to seven-fold increase in the nuclear-to-cytoplasmic ratio of both subunits was observed, suggesting engagement in the transcriptional alteration of target genes (Yang et al., 2004).

The activity of the entire GABP complex can be regulated by changing the redox status of GABP•, more precisely of its cysteine residues. Tipping the balance towards their oxidated form in expense of the reduced prevents heterodimerization with GABP• and causes impaired DNA binding, which reduces the total GABP activity (Martin et al., 1996).

Introduction

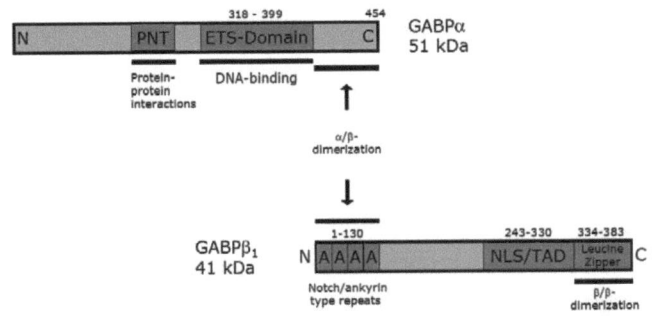

Fig. 1.2: **Schematic structure of GABP subunits**

GABP• bears a pointed domain (PNT) situated in the N-terminal part and an *Ets* DNA binding domain (DBD) in the carboxy terminal part. **GABP•** contains an N-terminal ankyrin (A) repeat domain, which mediates its interaction with GABP•, transcription activation domain (TAD) and nuclear localization signal (NLS), responsible for the targeting of the whole complex to the nucleus. Some isoforms (GABP•$_1$-41, GABP•$_1$-42 and GABP•$_2$) incorporate a C-terminally situated leucine-zipper domain (LZ), mediating the homodimerization of GABP• subunits.

1.2.2. The GABP• Subunit

GABP• is homologous to the *D. melanogaster* D-Elg factor, which is important for egg chamber patterning and oocyte development (Gajewski et al., 1995). GABP• is able to bind to its DNA recognition sites and necessary to recruit GABP• to target genes. The *Ets* DBD is near the carboxy terminus of GABP• and contains tryptophan repeats, forming the typical motif of *Ets* factors (Sharrocks, 2001). The Ets domain of GABP• is responsible for DNA binding, which was shown by mutational and deletion analysis (Thompson et al., 1991; LaMarco et al., 1991; Watanabe et al., 1990 and Gugneja et al., 1995). B. Graves (1998) compactly described its structure as follows: "In GABP•, helix 3 of the helix-turn-helix motif binds DNA in the major groove with two invariant arginines hydrogen bonding directly to the guanine residues of the (GGAA) core. Other structural elements, including the • sheet and helix 1 make phosphate contacts on each flank of the (GGAA) core sequence and these contacts indirectly specify additional sequence preferences" (see Fig. 1.1). The different affinity and preference of GABP• and other related transcription factors to the same binding site can be demonstrated by comparison with another *Ets* factor - PU.1. It forms phosphate bonds with DNA, which are similar, although not identical to those described for the crystal structure of GABP• (Pierson and Kennedy). In addition, the structures of both transcription factor-DNA bound complexes exhibit slightly different binding schemes between helix 3 and the GGAA motif. Although GABP and PU.1 bind to the same GGAA motif, PU.1 is known to have a higher affinity to an alternative core, AGA. As stated above, these variations can be considered as a direct consequence from the structural differences of the DNA-protein interactions (Graves 1998; Batchelor et al., 1998).

Thompson et al. (1991) proposed that GABP recognizes two GGA motifs that are separated by one-half turn of the DNA helix in the ...GGANNNGGA... motif. This spacing places the major groove interactions of the two ETS domains on opposite faces of the DNA helix. This orientation could accommodate a 9-bp contact zone for each ETS domain, as only 5 bp are proposed to be contacted in the major groove. The 3 bp that lie between the two GGA motifs would be recognized in the minor groove by the first ETS domain and in the major groove by the second ETS domain. Nevertheless, other data indicate the existence of significantly longer distances between the individual GABP-binding sites. Virbasius and colleagues observed that the distance between two GABP binding sites in tandem configuration (in their publication they used the name NRF-2 – see Table 1.1) can be between 20 and 30 bp without damaging the ability of GABP to bind to these sites at the same time (Virbasius et al., 1993). There are no known previous examples where GABP recognition site represents an inverted instead of a tandem repeat. Generally, tandem sites show a higher affinity than single sites. Examples for high affinity binding sites *in vitro* are those from the *IL2* distal promoter (Avots et al., 1997) and the *TSC-2* promoter (Ikeda et al., 2000). The affinity of binding does not always correlate with the strength of the effect on target gene expression. Weak binding sites can also be very important for the regulation of GABP target genes (Yang et al., 2007).

The region situated C-terminally from the DBD is necessary for the dimerization with the GABP• subunit. This is confirmed by crystallography analyses resolving the structure of bound fragments from GABP• , GABP• and the corresponding DNA binding region, the so-called "ternary complex" of binding molecules (Batchelor et al., 1998). Notably, while GABP•/• heterodimers can be formed spontaneously in solution, heterotetrameric complexes are formed only in the presence of DNA (Chinenov et al., 2000). Nevertheless, only GABP• and no other *Ets* factor is able to recruit GABP• to DNA (Brown et al., 1992; Ghosh et al., 1998).

The middle-region of GABP• contains the pointed domain, which can be found also in other *Ets* factors, where it aids important inter–protein connections (McLean et al., 1996). However, regarding the PNT domain of GABP• , no protein targets have been described yet. It cannot form spontaneously homodimers in solution (Chinenov et al., 2000), which makes it unlikely that this particular PNT domain can homodimerize. This assumption was confirmed by observations of Jousset et al. (1997). Nevertheless, it was proven that GABP• binds the transcriptional co-activator p300 (Bush et al., 2003) which suggests that the PNT domain may facilitate this physical interaction.

The human *GABPA* gene (encoding GABP•) was mapped to the long arm of chromosome 21 (HSA21) by fluorescence *in situ* hybridization (FISH) (Sawada et al., 1995), whereas the mouse gene, *Gabpa*, is located on chromosome 16 (de la Brousse et al., 1994). The long arms of both chromosomes – human and mouse (HSA21 and MMU16) share synteny for more than 100 genes (Reeves and Cabin, 1999), including *Gabpa*. Ten exons for the *GABPA* gene were identified accounting for the entire cDNA sequence (Goto et al., 1995).

The transcription of the *GABPA* gene is linked to that of the ATP Synthase Coupling Factor 6 (CF6) gene (*ATP5J*) and is controlled by a common bidirectional promoter (Chinenov et al, 2000). This promoter contains four GABP binding sites, one Spl/3 binding site and one YY1 binding site. The GABP-1 site located proximal to the transcription start sites functions cooperatively with the other GABP binding sites and with the Spl/3 and YY1 sites and activates the GABP• and CF6 promoters. Binding of GABP to the GABP• /CF6 bi-directional promoter provides the potential for autoregulation of GABP• expression and indicates the importance of GABP in the coordinate expression of respiratory chain components (Patton et al., 2006).

In human Down syndrome (DS) fibroblast cell lines the chromosome 21 trisomy causes overexpression of GABP• mRNA. In spite of it, neither in these cells nor in GABP• -transfected NIH-3T3 cells differences in protein levels could be found (O'Leary et al., 2004). In addition, in the brain and in skeletal muscle of Ts65Dn segmental trisomy mouse model of Down Syndrome (DS), where the *Gabpa* gene dosage is increased, GABP• protein levels are also elevated. This suggests that although GABP• can be overexpressed, its protein levels are tightly regulated, probably in a tissue-specific manner. Therefore, GABP could be potentially involved in DS pathologies in those tissues where GABP• protein levels are indeed higher.

GABP• expression can be observed in all cell types. Using Northern analysis and immunoblotting it has been detected in hematopoietic cells and various organs, as testis, muscle, liver etc. (LaMarco et al., 1991 and Brown et al., 1992). Single GABP• transcripts have been characterized in both human and mouse cells (Chrast et al., 1995). In mouse cells, however, *GABPA* has been described to be alternatively spliced. The tissue-specific exons 1 is combined with four forms of 3' polyadenylation signals, which all together result in twelve specific transcripts for *GABPA* (O'Leary et al., 2005). These different transcripts are suggested to alter stability, sub-cellular localization or possibly translation efficiency. A processed GABP• pseudogene was identified in humans and mice, displaying more than 97% homology to the genuine *GABPA/Gabpa* gene. In myeloid cells, RNA from this pseudogene is transcribed, but not translated, which is prevented by a mutation, occurring in the translation initiation codon (ATG) (Lue et al., 1999). Hypothetically, this pseudogene originated when an ancient retrovirus, which captured a mutated GABP• transcript, reinserted back into the genome.

1.2.3. The GABP• Subunit

In mouse two distinct GABP• genes are located on chromosome 2 (*Gabpb1*) and on chromosome 3 (*Gabpb2*) (de la Brousse et al., 1994). Furthermore, GABP•$_1$ is expressed in four distinguishable isoforms, resulting from alternative splicing of mRNA (Gugneja et al., 1995). Due to the large variety of isoforms, the GABP nomenclature becomes complex and sometimes confusing. Nomenclature, naming the GABP•• isoforms was proposed by Rosmarin et al., 2004, based on their particular molecular weights (See Table 1). GABP•$_2$ is less well characterized, and it is not known if alternative isoforms are expressed (de la Brousse et al., 1994).

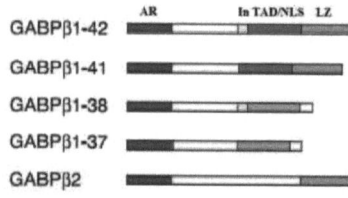

Fig. 1.3. Comparison of the GABP• family members
(Rosmarin et al., 2004 - modified)

The GABP• ankyrin repeat domain, mediating interaction with GABP• is localized at the N-terminal region and is homologous in all the isoforms of this subunit. In contrary, in the central region of this protein is situated a 12-amino-acid region (In), which is present only in two of the isoforms (GABP•-42 and GABP•-38), has unknown function and contains two pairs of serine residues. Differences among the GABP• isoforms are observed also in their carboxy termini, containing the TAD (transcription activating domain), NLS (nuclear localization signal) and LZ (leucine zipper) domain. Contradictory data exist concerning the exact localization and structure of TAD and NLS.

All GABP•$_1$ isoforms (Fig. 1.3) utilize a shared amino terminus, containing four and a half ankyrin repeats. This protein structure is present also in several other proteins, such as Notch, InB and p19Arf and functions as a mediator of inter-protein connections (Sedgwick and Smerdon, 1999). The ankyrin motifs in GABP• are required for its recruitment to the DNA, since they bind to the C-terminus of GABP• and thus help to

Introduction

assemble the complete triplex DNA-GABP•-GABP• (Thompson et al., 1991; Gugneja et al., 1995 and Sawa et al., 1996). Ternary complexes between GABP•, GABP• and other *Ets* proteins could not be found, which demonstrates the high degree of specificity of GABP• for GABP• (Brown and McKnight, 1992). Earlier structure–function analyses were confirmed by studies on the crystal structure, which visualize the physical contact between the GABP• C-terminus and GABP• occurring through the ankyrin repeats (Batchelor et al., 1998).

The separate GABP•$_1$ molecules homodimerize via their carboxy termini. Each C-terminus contains a domain, rich in hydrophobic residues, possessing high binding affinity to similar structures. These domains assume helical conformations and tend to interweave with each other as spiraling loops. This mode of action bears resemblance to a zipper, which is reflected in their name: Leucine Zipper (LZ). The LZ domain is present only in the longer isoforms GABP•$_1$-42 and GABP•$_1$-41 and also in GABP•$_2$, which is able to heterodimerize with the long GABP•$_1$ isoforms (de la Brousse et al., 1994). The described ability of GABP• subunits to dimerize is obligatory for the formation of GABP heterotetrameric complex - $(••)_2$. The flexible links along the whole span of GABP• molecule – from the leucine zipper at the C-terminus to the ankyrin repeats at the N-terminus – is thought to be the reason for the ability of GABP to recognize direct and inverted repeats of the GGAA core sequence with variable spacing (Fig. 1.1).

Other differences between the isoforms of GABP•$_1$ are due to the presence or absence of a 12-amino-acid section in the central region of the protein (Gugneja et al., 1996). Not much is known about this domain, except that it is present only in GABP•$_1$-42 and GABP•.-38 and possesses two neighbouring pairs of serine residues. As GABP• is known to undergo phosphorylation by serine-threonine kinases (Flory et al., 1996), the presence of these clustered serine residues raises the possibility that they may serve as a phosphorylation site (Gugneja et al., 1995).

Contradictory data are available about the location and structure of the transcription activation domain within GABP•. There is general agreement that it is situated in the carboxy terminal half of GABP•$_1$. The exact position and span of the TAD however, as well as the ability for transcriptional activation of the different isoforms is a matter of debate. Gugneja and colleagues determined that the TAD lies between residues 258 and 327 of GABP•$_1$. This position puts it in a region which is common to all GABP•$_1$ isoforms. Their more precise mutagenesis experiments defined glutamine-containing hydrophobic structures within this domain, upon which depends the

Introduction

transcriptional activity (Gugneja et al., 1995). Using fusion protein constructs between all GABP·$_1$ isoforms and the DBD from the yeast Gal4 transcription factor, they co-transfected them into COS cells together with an artificial reporter under the control of five copies of a Gal4 DNA binding site. The results showed that all four fusion isoforms strongly transactivate the reporter gene up to the same level (Gugneja et al., 1995). Interestingly, the additional deletions that were made at the very end of the carboxy terminus resulted in an even stronger increase in the transcriptional activation. A conclusion was drawn that all four GABP·$_1$ isoforms share a common glutamine-rich TAD.

Surprisingly, another group, Sawa and colleagues identified the location of TAD at the very end of the carboxy terminus of GABP·$_1$ – area that differs among the various isoforms. Using mutational studies, they reasoned that the NLS of hGABP·$_1$-41 and hGABP·$_1$-37 is positioned between the amino acid sequences 243–330, which is not overlapping with the TAD (Sawa et al., 1996). They found that while GABP·$_1$-42 and GABP·$_1$-41 were transcriptionally active, the GABP·$_1$-38 and GABP·$_1$-37 isoforms were transcriptionally silent. Their conclusion was that the TAD was inseparable from the GABP·$_1$ homodimerization region at the very carboxy terminus of the molecules (Sawa et al., 1996).

Such incompatible results can be explained by the differences in methodology, used by the two groups. Gugneja et al. used Gal4-GABP· fusion proteins, together with an artificial reporter. Thus, any involvement of GABP· was effectively excluded. It is worth to mention that their deletion analysis destroyed the original GABP· NLS, and the NLS was provided by Gal4 DBD.

Sawa and colleagues used a more natural template, but it was introduced into *Drosophila* cells which lack endogeneous GABP. Therefore, each approach had its own specific properties contributing to the observed contradictory results.

Rosmarin et al., 2004 published their own data concerning the same question. They tried to avoid the flaws of the previous attempts, by using "intact transcription factor molecules on a natural target gene promoter in a proper cellular context" (U937 myeloid cell line and a luciferase reporter, joined to the GABP target gene *ITGB2* (CD18-gene)). After co-transfection of the different GABP· isoforms with GABP· and the reporter construct, they found that GABP·$_1$-42 and GABP·$_1$-41 were 2-fold more transcriptionally active than GABP·$_1$-38 and GABP·$_1$-37, which partially confirms the results by Sawa et al., 1996. However, in contrast to Sawa et al., Rosmarin and colleagues found significant transcriptional activity in the latter two isoforms, while Sawa et al. found none.

Surprisingly, during the differentiation of myeloid cells, an interesting expression pattern of the GABP• isoforms was found (Rosmarin et al., 2004). In the process of differentiation of granulocytes, the expression of isoforms, excluding the 12-amino-acid insert (GABP•$_1$-41 and GABP•$_1$-37) raises, while the expression of other isoforms diminishes. In contrast, during monocytic differentiation, a similar discriminative expression of GABP• isoforms has not been detected. When differential expression of this isoforms takes part, it seems to be regulated at the RNA splicing stage, because the same pattern is seen both at transcription and at expression levels. Although only subtle 2-fold differences in activity were observed, the differential expression of GABP• isoforms eventually plays a role in regulating gene expression also in other cells, due to their different abilities for activating the expression of target genes. Later on, nine splice variants of *Gabpb1* were identified in mice (O'Leary et al., 2005). These differentially expressed forms lack functional domains, supposing a dominant negative function.

1.3. GABP• •• Target Genes – from *in vitro* to *in vivo*

More than 60 potential GABP target genes have been identified by *in vitro* experiments, including both lineage-restricted and housekeeping genes. Several studies have demonstrated that GABP-binding sites are present in many TATA-less promoters, where it might initiate the assembly of the pre-initiation complex (Ogawa et al., 1988; Boone et al., 1990; Gutman et al., 1990; Huhtala et al., 1990; Campbell et al., 1991; Bottinger et al., 1994 and Villena et al., 1994). A group of authors suggested a model of action for the GABP function, according to which GABP binds to its recognition DNA site and acts as an assembly platform for the transcription initiation apparatus, containing Sp1, RNA polymerase II and TFIID (Yu et al., 1997). This mechanism might represent a specific class of promoters, characterized with the ability and necessity to synchronously bind different Ets factors. As an example could be given the minimal CD18 promoter, which lacks a TATA box, CAAT and initiator elements, and is composed mainly of Ets repeats. The binding Ets factors may act via direct recruitment of elements from the main transcription system, thus starting the transcriptional activation. Supposedly, such binding - coordinated by repeated elements - is thought to be a normal feature for some TATA-less promoters, especially for those that direct spacial or temporal-specific expression by multiple transcription starting sites (Bottinger et al., 1994). The individual GABP binding

Introduction

sites, however, may be organized in different manner for every particular promoter. The distance between GABP-binding tandem repeats, for example, may vary between 2 bp (Thompson et al., 1991) and 20 -30 bp apart without damaging the ability of GABP to bind to these sites at the same time (Virbasius et al., 1993) (For details see chapter 1.2.2).

Regardless of its wide-spread expression, GABP also modulates the function of many lineage-specific genes, as exemplified in myeloid cell lineage where GABP regulates CD18 (Rosmarin et al., 2005 and Bottinger et al., 1994), •• integrin (Rosen et al., 1994), lysozyme (Nickel et al., 1995), neutrophil elastase (Nuchprayoon et al., 1997) and folate receptor•• ••genes (Sadasivan et al., 1994).

The leukocyte •2 integrin CD18 plays important roles in the defense of the organism against pathogens. During the differentiation of myeloid precursor cells to granulocytes and monocytes the expression of CD18 increases (Rosmarin et al, 1989 and Hickstein et al., 1988), which coincides with the pattern of increase observed for some of the GABP isoforms during the same differentiation process (see above). This is most probably due to the fact that CD18 transcription can be activated by GABP which binds to its promoter (Rosmarin et al., 2005 and Bottinger et al., 1994) and works in cooperation with the transcriptional activator Sp1 (Rosmarin et al., 1998) and with another *Ets* factor, PU.1 (Rosmarin et al., 1995).

The joint action between these three transcription factors is controlling the response of CD18's promoter to retinoic acid (RA). In myeloid cells, RA stimulates the process of differentiation to mature granulocytes (Gaines and Berliner, 2003) and CD18's transcription is activated by RA (Hickstein et al., 1988 and Rosmarin et al, 1989). RA binds to the retinoic acid receptors (RARs), triggering a chain of processes resulting in transcriptional activation of the target genes. In CD18's gene promoter exists a distal enhancer element, able to bind RARs, but it is responsible only partially for the RA sensitivity of CD18's gene expression. Transfecting a set of CD18 promoter constructs into myeloid cells, Alan G. Rosmarin and colleagues discovered that only a half of the RA responsiveness is dependent on the proximal promoter (Rosmarin et al., 2004). It does not bind RARs but binds Sp1, PU.1 and GABP instead. Mutational analysis demonstrated that RA responsiveness requires the presence of Sp1- and GABP-binding sites. This is proven by Bush et al. (2003), who disrupted these sites and, thereby, strongly reduced the RA responsiveness of CD18 gene (*ITGB2*). After the discovery that GABP and Sp1 cooperate to activate the folate receptor • (Sadasivan et al., 1994) it became clear that a similar mechanism is utilized for mediating the RA responsiveness of its gene (Hao et al., 2003). In an analogous way, the hormone responsiveness of the HIV LTR (Rahman et al., 1995) and of the prolactin promoter is mediated (Ouyang et al., 1996 and Schweppe and Gutierrez-Hartmann, 2001). This demonstrates how GABP in cooperation with other widely expressed transcription factors integrates intracellular signals and mediates the hormone responsiveness of some very specific, lineage-restricted genes. In addition, GABP together with SP1-plays a role in controlling the expression of *Robo4*. This is a member of the roundabout receptor family that is expressed only in endothelial cells. It is important for the endothelial cell migration and involved in angiogenesis. When in primary human endothelial cells the amounts of GABP and SP1 are knocked down by transfection with siRNA, a 50% reduction in the normal levels of endogenous Robo4 mRNA expression is observed (Okada et al., 2007).

Furthermore, it has been demonstrated that transactivation of the blood coagulation factor IX promoter depends on the binding of GABP• •• complex to the so-called "site 5". The binding of C/EBP• to the same site also augments this activity, although in supershift experiments GABP comprises the majority (approximately 50%) of the total site 5 binding activity (Boccia et al., 1996).

In lymphocytes GABP increases transcription from the distal enhancer of the *IL2* gene (Avots et al., 1997), from the Fas/CD95 promoter (Li et al., 1999) and from the interleukin 16 promoter (Bannert et al., 1999). It is also a critical regulator of B lymphocyte development, probably influencing the expression of Pax5 and its target genes. One of them is *Cd79a*, which is strongly downregulated in GABP• -deficient B cells and their progenitors. This notion is supported by the fact that *in vivo* GABP is able to bind to regulatory regions both in the Pax5 and Cd79a promoters (Xue et al., 2007).

In mast cells, GABP governs the synthesis of heparin by directing the expression of the crucial enzyme NDST-2 (N-deacetylase/N-sulfotransferase 2) (Morii et al., 2001). GABP plays a significant role in the formation of skeletal muscles and the establishment of neuromuscular synapses (Briguet and Ruegg, 2000). It influences the transcription of AChR (nicotinic acetylcholine receptor) subunits • and • (Schaeffer et al., 1998 and Fromm and Burden, 1998) and the muscle-specific utrophin genes (Koike et al., 1995; Duclert et al., 1996; Khurana et al., 1999; Gramolini et al., 1999; Galvagni et al., 2001 and Gyrd-Hansen et al., 2002).

Modulation of uthropin expression could be the basis for a therapy of Duchenne muscular dystrophy (DMD) (Khurana and Davies, 2003). Neuregulin-stimulated phosphorylation of PGC-1• and GABP provokes their conjugation into one macromolecular complex, which functions as an enhancer of the transcription of a broad gene expression program, specific for the myoneural synapse. Since a number of genes from this expression program are deregulated in DMD, the PGC-1• -involvement in this process was tested by expressing it in a muscle of *mdx* mice, used as a disease model (Handschin et al., 2007). Such animals show improvements in the properties, characteristic for DMD, namely in muscle histology, running performance, etc. Thus, the control of PGC-1• levels and, hence, of GABP activity in skeletal muscle is a potent candidate for prevention or treatment of DMD. In connection with the subject is reported that conditional gene targeting of *Gabpa* disrupts the synaptic function of neuromuscular junction by disrupting the regulation of postsynaptic genes expression (O'Leary et al., 2007). In skeletal muscle tissues of conditional GABP• knockout mice reduction in the expression of the acetylcholine receptor • subunit was proven, as well as an increase in the expression of the • subunit. Interestingly, another group obtained different results, claiming that GABP is neither necessary for neuromuscular synapse formation nor for synapse-specific gene expression (Jaworski et al., 2007). They state that the AChR • expression level is independent upon GABP. Differences in the experimental design and methodology, like leaving the expression of partial GABP• transcripts and incomplete knockout of the protein, might account for the divergent results.

A variety of vital housekeeping genes is also regulated by GABP. For instance, it controls the expression of several nuclear genes, encoding key components from the respiratory chain. In fact, Virbasius and colleagues identified GABP exactly in this setting, accordingly giving it the name Nuclear Respiratory Factor 2 or NRF-2 (Virbasius et al., 1993). It was proven *in vitro* that the mammalian GABP complex, together with structurally unrelated transcription factor NRF-1 regulates the expression of a variety of housekeeping genes, including many responsible for the cellular respiration in mitochondria (Scarpulla et al., 2002). Among them are cytochrome C oxidase (COX) subunits IV, Vb, VIIa (Virbasius et al., 1993; Carter and Avadhani, 1994; Sucharov et al., 1995; Martin et al., 1996 and Seelan et al., 1996), VIA1 (Wong-Riley et al., 2000), VIIAL (Seelan et al. 1996), VIIC (Seelan et al. 1997) and XVII (Takahashi et al., 2002). It also includes mitochondrial transcription factor A (MTFA) (Virbasius et al., 1994), the main transcription factor, involved in the control of mitochondrial gene expression, which also activates the replication of mitochondrial DNA, and the ATP synthase • gene (Villena et al, 1998). Therefore, GABP can be termed as a "central regulator" of the energy metabolism.

GABP is thought to play an important role in the coordinated expression of oxidative phosphorylation (*OXPHOS*) genes, including cytochrome oxidase subunit genes associated with mitochondrial biogenesis in brown adipocyte differentiation (Villena et al., 1998). Estrogen-related receptor • (Err•) and GABP were identified as key transcription factors regulating the *OXPHOS* pathway in human diabetic and pre-diabetic skeletal muscle (Mootha et al., 2004).

One more fact about the GABP involvement in cell differentiation is presented by Day et al. (2004) who demonstrated that expression of both GABP• and -• subunits is increased during the differentiation of adherent peripheral blood mononuclear cells into the multinucleated, resembling osteoclasts cells in the presence of receptor activator of NF-kB ligand and macrophage colony stimulating factor (M-CSF). Importantly, addition of M-CSF only, provoked the development of macrophage-like instead of multinucleate osteoclast-like cells and did not result in an increase of GABP• •• expression levels.

GABP is a proposed regulator of expression of the ribosomal proteins L27A, L30, L32 and S16 (Curcić et al., 1997 and Genuario et al., 1993), indicating its essential function in the energy metabolism and, possibly in protein synthesis (Perry R. P., 2005). In this case, complete inactivation of GABP in all cell types may be incompatible with the survival of a targeted organism.

Strong evidence supporting this assumption was given by A. Avots, 2002 (unpublished data) and Ristevski et al., 2004 who created GABP• knockout mice and observed lethality at very early stages of embryonic development (prior implantation) for $GABP^{-/-}$ embryos. This fact is consistent with previous observations that GABP• is widely expressed in embryonic stem cells and during embryogenesis suggesting an important function for GABP• during those stages. *Gabp•* heterozygous mice, however, manifest no detectable phenotype and show unchanged protein levels in all tissues examined. This shows that GABP• protein levels are strictly regulated and indicates that GABP• function is essential and can not be compensated by other *Ets* transcription factors during development. Additionally, the expression level of Oct-3/4 (a transcription factor which is essential for the self-renewal of Embryonic Stem (ES) cells) was reduced by knockdown of GABP• expression (Kinoshita et al., 2007). It was found that GABP upregulates the expression of Oct-3/4 via down-regulation of Oct-3/4 repressors indicating an important role of GABP in the processes of early embryogenesis.

Introduction

GABP also affects other important cellular functions, such as the cell cycle control (Fig. 1.4). One target is the retinoblastoma gene (*RB1*), which is controlled transcriptionally by GABP (Savoysky et al., 1994; Shiio et al., 1996 and Sowa et al., 1997) and whose product binds to the transcription factor E2F1, thus effectively sequestering it at the G_1/S restriction point. Interestingly, the expression of DNA polymerase • (Izumi et al., 2000 and Yang et al, 2007) and thymidylate synthase (Rudge and Johnson, 2002 and Yang et al, 2007) is influenced by both E2F1 and GABP. E2F1 is able to interact physically and functionally with GABP•$_1$, one of the components of the GABP complex (Hauck et al., 2002). GABP is also essential for the transcription of Skp2 (S-phase Kinase-associated Protein 2), a member of the F-box protein family. Its binding to the corresponding recognition site within the *Skp2* promoter depends on the cell cycle and regulates *Skp2* expression at the restriction point from G_1 to S phase (Imaki et al., 2003). The authors observed that the overexpression of GABP• and the suppression of GABP• or GABP• by a siRNA results in the increase and respectively reduction of Skp2 promoter activity.

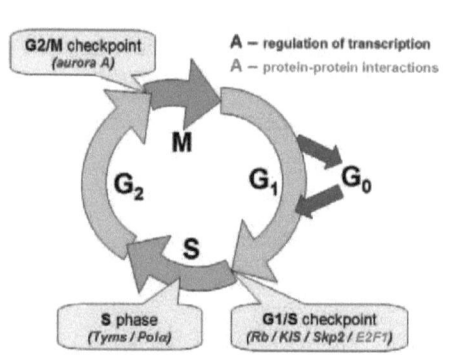

Fig. 1.4: **Involvement of GABP in the cell cycle regulation.**

Genetic disruption of *Gabpa* is abolishing entry into S phase. Moreover, the expression of genes required for DNA synthesis and for degradation of cyclin-dependent kinase inhibitors is selectively reduced, such as the thymidylate synthase (*Tyms*) gene and the gene encoding DNA polymerase • p180 catalytic subunit (*Pola*) (Yang et al., 2007). GABP relieves repression from the CDKIs p21 and p27 by stimulating the expression of Skp2, which targets and degrades them. Skp2 is upregulated specifically in the period of late G_1–S phase. The amounts of Skp2 protein are in equilibrium between their GABP-induced expression and Apc-controlled protein degradation. Therefore, Skp2

Introduction

contributes in the cell-cycle arrest in *Gabpa* double negative cells through degrading CDKIs. Furthermore, well-designed experiments indicated that ectopic expression of both GABP• and —• is required and sufficient for quiescent cells to re-enter into the cell cycle through a pathway other than that of D-type cyclins and CDKs (Yang et al., 2007).

GABP also indirectly regulates the cyclin-dependent kinase inhibitor p27 (Kip1) by phosphorylating its serine 10 residue by the protein kinase KIS, leading to nuclear export of Kip1 and protein degradation. The promoter region of KIS incorporates two Ets-binding sites, binding 3 or more specific complexes containing GABP. Both sites are equally important, as mutation in either of them leads to an estimated 75-80% reduction in promoter activity. Decreasing the amounts of GABP• by introducing siRNA in vascular smooth muscle cells lowers KIS gene (*UHMK1*) expression, which in turn is associated with an increase in Kip1 protein levels and, respectively, with smaller percentage of cells in S-phase (Crook et al., 2007). All these examples demonstrate that GABP regulates the expression of crucial components of the cell cycle and, in particular, the G_1/S restriction point.

In respect to the G_2/M checkpoint, GABP was identified as a positive regulator of *aurora A* transcription (Tanaka et al., 2002). Aurora A is an important serine-threonine kinase, controlling several mitotic events. Its transcription is altered throughout the cell cycle, with a peak during G_2/M-transition. Co-expression of GABP• and GABP• markedly increased *aurora A* promoter activity.

Transcription of several key genes from important human viruses, including the HIV-1 LTR (Verhoef et al., 1999), *ICP0*, *ICP4* and IE genes of herpes simplex virus (Douville et al., 1995 and Vogel and Kristie, 2000) and the *E4* gene of adenoviruses (Watanabe et al., 1988) at least partially depends on GABP as a host cell factor. This suggests that the utilization of GABP by viruses to achieve high-level expression is caused by its powerful transcriptional properties.

In summary, the observations listed in this chapter show that GABP is critical for several physiological processes. It transcriptionally regulates basic cellular functions, such as energy metabolism, protein synthesis and cell cycle. In addition, GABP is involved in human pathophysiology, including infection by important viral pathogens. Thus, it is anticipated that conventional gene knockout is incompatible with survival of the organism and with the normal development of the cell.

1.4. GABP in Signalling Processes

During differentiation, maturation and activation of hemato-lymphoid cells, as an important transcriptional regulator of key cytokines and their receptors, GABP is a direct signalling target of various pathways. For instance, GABP targets the distal *IL2* enhancer in lymphocytes and contributes in IL-2 induction (Avots et al., 1997). GABP also transcriptionally controls another important cytokine, i.e. IL-16 in T cells (Bannert et al., 1999) and regulates the thrombopoietin (*TPHO*) gene, responsible for megakaryocyte development (Kamura et al., 1997). GABP also binds to an enhancer site in the third intron of the proinflammatory cytokine tumor necrosis factor • (*Tnf*) gene and regulates its expression (Tomaras et al., 1999). GABP also controls the expression of many transmembrane cytokine receptors by binding to and activating the •c chain promoter, which is one of the components of the cytokine receptors for IL-2, IL-4, IL-7, IL-9 and IL-15. Defects in •c or its expression cause a severe combined immunodeficiency (Markiewicz et al., 1996). The IL-7 receptor • -chain (IL-7R•) is essential for B- and T-cell development and is a component for the receptor of thymic stromal lymphopoietin. In humans, mutation of IL-7R• results in a selective loss of T cells (Puel et al., 1998). Whereas the lineage-specific Ets-factor PU.1 regulates IL-7• expression in mouse pro-B cells, by using GABP• deficient mice it was demonstrated that GABP is essential for the regulation of IL-7• expression in T cells where PU.1 is not expressed (Xue et al., 2004). These examples illustrate that GABP regulates cytokines and their receptors, vital to the immune response and inflammatory processes and crucial for platelet growth, differentiation and maturation.

An additional group of target genes which are controlled by GABP are several genes encoding hormones and their receptors, mediating mammalian reproductive functions and controlling the thyroid gland. One of them is the oxytocin receptor (OTR), credited for the initiation of milk ejection. Its promoter is activated only after binding of both GABP together with c-Fos/c-Jun (AP-1) (Hoare et al., 1999). Other targets of GABP are the prohormone prolactin which is involved in the regulation of reproduction, including the mammary gland development (Ouyang et al., 1996 and Schweppe and Gutierrez-Hartmann, 2001), and the thyrotropin receptor gene which is involved in thyroid gland growth and development (Yokomori et al., 1998).

Several signalling pathways are targeting the GABP• •• complex. Activation of T lymphocytes leads to the induction of *IL2* gene, and GABP regulates its distal enhancer (Avots et al., 1997). During T cell activation mitogen-activated ERK and SAPK protein kinase cascades are stimulated. One member of these cascades, Raf-1 kinase, can induce *in vivo* the phosphorylation of GABP, although via an indirect mechanism. *In vitro*, neither of the GABP subunits could be phosphorylated with purified Raf-1 or MEK. Instead, this could be done only by reconstituting the MAPK/ERK cascade suggesting that GABP is a substrate of this pathway. (Flory et al., 1996) Hence, GABP is not a direct target of Raf-1 kinase but of classical cytoplasmic MAP/Erk kinase cascade. Both subunits of GABP are also phosphorylated upon JNK activation *in vivo* (Hoffmeyer et al., 1998). In addition, upon phosphorylation, GABP causes an increase in the activity of *IL2* distal enhancer (Avots et al., 1997 and Hoffmeyer et al., 1998), which all together indicates that MAP/Erk signalling pathways are involved in the IL-2 transcription in T lymphocytes.

This is not the only example for signalling pathways which control the transcriptional activation of GABP. Upon GABP binding to the prolactin promoter and in presence of insulin, prolactin transcription is activated. Insulin treatment of pituitary cells is elevating the levels of MAPK activity and results in phosphorylation of GABP•• but not of GABP•. Therefore, GABP may be regulated also by MAP kinase phosphorylation under these circumstances (Ouyang et al., 1996).

Likewise, during neuromuscular synapse formation, GABP binding to the AChR promoter is necessary for neuregulin-1 (NRG-1) to transcriptionally upregulate the acetylcholine receptor (Schaeffer et al., 1998; Fromm and Burden, 1998). NRG-1 signalling stimulates transcription not directly, but by specifically increasing only the transcriptional activity of GABP. However, this is not affecting its DNA-binding activity (Fromm and Rhode, 2004 and Sunesen et al., 2003). This might be due to phosphorylation of GABP, supported by the observation that in *in vitro* experiments both GABP subunits can be phosphorylated by Erk and JNK, while NRG-1 is able to stimulate GABP• phosphorylation *in vivo* (Fromm and Burden, 2001). In a very similar manner, NRG-1 is using GABP for increasing the expression of the utrophin (*Utrn/UTRN*) – gene in muscle cells, suggesting that GABP• is target of MAP kinases in these cells and its phosphorylation is causing the transcriptional activation of both AChR and utrophin genes by NRG-1 (Gramolini et al., 1999 and Khurana et al., 1999).

1.5. GABP Protein–Protein Interactions – Partner Proteins

The expression of one large group of GABP target genes is specifically regulated either by spatial tissue-restriction or transitionally – only during cellular differentiation and/or activation. Nevertheless, their pattern of expression does not correlate well with the almost ubiquitous expression levels of GABP•••. One of the mechanisms regulating the expression of such target genes might be the tissue-specific phosphorylation of GABP, which takes place in response to different cellular signalling events.

As already discussed above, the most crucial protein–protein interactions of the GABP••• hetero-tetramer are those occurring between the individual subunits of the complex itself, required for the combination of DNA-binding and transcriptional activity (see chapter 1.2.). The second very important mechanism for regulating the lineage-restricted target genes that GABP is utilizing is via the combined actions of several transcription factors. Such as the regulation of CD18 expression in which GABP cooperates with PU.1 and Sp1 (Rosmarin et al., 1995, 1995 and 1998), and the neutrophil elastase expression in myeloid cells in which GABP, C/EBP•, PU.1 and c-Myb are needed for proper regulation (Nuchprayoon et al., 1997). Hence, physical interactions between these factors have been described. These examples show a cooperative activity of GABP together with multiple other transcription factors. Non of which is truly tissue-specific, but nevertheless they altogether create a mechanism for the specific control of lineage-restricted genes.

1.5.1. Interactions with GABP•
1.5.1.1. Sp1

Sp1 is a transcription factor, which is able to bind to numerous TATA-less promoters and whose DNA binding site is rich of GC nucleotides. It interacts directly with molecules from the basal transcriptional system and is reported to cooperate functionally with GABP in the activation of some target genes, both in the control of lineage specific and widely expressed genes. As examples, the genes encoding CD18 (Rosmarin et al., 1998), utrophin (Galvagni et al., 2001 and Gyrd-Hansen et al., 2002), heparanase-1 (Jiang et al., 2002), the *Pem Pd* homeobox gene (Rao et al., 2002), the folate receptor • (Sadasivan et al., 1994) and human sulfotransferase SULT1A1 (Hempel et al., 2004) can be given. In each of these cases, interactions amongst GABP and Sp1 have been described, which could be illustrated for the TATA-less promoter of *SULT1A1*. Its high activity is dependent both on the Sp1 and Ets binding sites, the latter of which could be bound by GABP. Transfection with GABP-expressing vector alone is able to transactivate the promoter in *Drosophila* S2 cells, but Sp1, shown in co-transfections, is the one that synergistically raises the GABP-mediated activation as much as 10-fold. As a negative control the closely related highly homologous promoter of SULT1A3 can be taken, which has a two-basepair difference in the Ets binding site that prevents the synergistic interaction between the two factors. Hence, it displays 70% lower activity, in spite of the fact that Sp1 and GABP alone can also induce the *SULT1A3* promoter (Hempel et al., 2004). In addition, it a direct physical interaction was shown between the zinc finger DNA binding domain of Sp1 and GABP• (Galvagni et al., 2001).

1.5.1.2. PU.1

The transcription factor PU.1 is a member of *Ets* family and important for myeloid cell development. It is also expressed in B-lymphocytes and in some monocytes. Abnormalities in its expression lead to embryonic lethality and severe defects in the corresponding cells (Scott et al., 1994 and Anderson et al., 1998). They can be linked with the development of Acute Myeloid Leukemia (AML) (Mueller et al., 2002).

GABP and PU.1 physically compete for attaching the same DNA binding sites within the •2 integrin CD18 promoter and, in fact, functionally cooperate to activate its transcription (Rosmarin et al., 1995). In HeLa cells, which normally do not express CD18, none of the transcription factors is sufficient on its own to activate CD18's promoter. Only the transfection of both of them together is sufficient for activation (Rosmarin et al., 2005).

Acting in a similar manner, GABP and PU.1 cooperate in the activation of the neutrophil elastase promoter, binding to its *Ets* sites (Nuchprayoon et al., 1997 and Oelgeschlager et al., 1996). Also, PU.1 and GABP, acting through the proximal and distal GGAA-elements are driving the transcription of the minimal defensin-1 (def1) promoter in HL-60 cells. Notably, PU.1 is dispensable for achieving the full effect. Creation of a strong TATA box in this promoter can substitute for the PU.1, but not for the GABP binding site (Yaneva et al., 2006). A direct physical interaction between these two transcription factors has not been described yet. However, both of them interact with the transcriptional coactivator p300/CBP (Bush et al., 2003; Bannert et al., 1999 and Blobel, 2000) which might represent the link required for functional interaction.

1.5.1.3. Activating Transcription Factor 1 (ATF1)

GABP, in synergy with ATF1 and CREB1, is activating the adenovirus early 4 (E4) promoter. Sawada and colleagues revealed that ATF1 and CREB1 interact only with GABP• , and not with GABP• (Sawada et al., 1999). Notably, from the whole ATF family, GABP• binds only to ATF1. The function of GABP• in the process is to strengthen the

GABP• –ATF/CREB physical interaction. The mechanism of achieving this may resemble the model with which GABP• changes the conformation of GABP•, thereby increasing its binding abilities. Another example for the functional interaction between GABP and CREB is the BRCA1 proximal promoter which is activated by both CREB and GABP (Atlas et al., 2000). The authors proposed that cooperation between these two factors might change the cAMP-responsiveness of the BRCA1 promoter.

1.5.1.4. p300/ CREB-binding protein (CBP)

p300 shares a high degree of homology with CBP. Both molecules are co-activators of many transcription factors and integrate various signals in cell-specific transcriptional responses (Vo and Goodman, 2001). Due to the nature of their function, they may assume the role of limiting factors in transcriptional activation, as suggested from gene disruption studies (Kung et al., 2000). Several *Ets*-transcription factors belong to the large group of transcription factors that physically and functionally interact with p300 (Blobel, 2000). For the first time interactions between GABP• (but not GABP•) and p300 were identified in T lymphocytes in which they regulate the IL-16 promoter (Bannert et al., 1999). p300 is also involved in the stimulation of another GABP-regulated gene, the gene encoding CD18, where it increases the RA responsiveness of the promoter through physical interaction with GABP• (Bush et al., 2003).

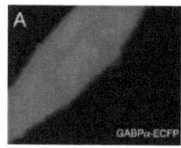
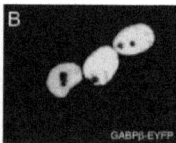
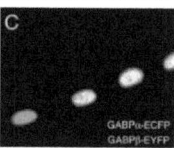

Fig. 1.5: Subcellular localization of GABP• and GABP•.
Epifluorescence microscope images, taken with the filter set given by the color in the panel. C2C12 myoblasts, transfected with: **A**, GABP•-N1-ECFP (the endogenous GABP• is not able to transport the excess of overexpressed exogenous GABP• into the nucleus and therefore it is left in the cytoplasm) or **B**, GABP•-C1-EYFP fusion proteins (the overexpressed exogenous GABP• effectively targets itself to the nucleus) or **C**, cotransfected with GABP•-N1-ECFP and GABP•-C1-EYFP [ECFP filter set] (the overexpressed exogenous GABP• is in sufficient amounts to bind to and to translocate overexpressed exogenous GABP• to nucleus) [adapted from (Sunesen et al., 2003)].

1.5.1.5. Microphthalmia-Associated Transcription Factor (MITF)

MITF is a leucine zipper protein forming a basic-helix-loop-helix structure, member of bHLH-ZIP family. MITF's role is better understood by examining a mutant allele – the *mi*-MITF, containing a deletion within an important region, which abolishes its most important functions – the nuclear localization and transcriptional activation. In mast cells this mutant allele is disturbing the *Ndst2* gene expression, which decreases the heparin levels (Morii et al., 2001). MITF does not directly regulate *Ndst2* gene though – the main transcription factor activating NDST is GABP. The mode of action, through which MITF enhances GABP function, is via facilitating its nuclear localization. GABP• is not able to enter the nucleus itself, instead it has to be transported there by a partner molecule, possessing Nuclear Localization Signal (NLS) domain. GABP• can bind to wild-type MITF, *mi*-MITF and GABP•, but only GABP• and wild-type MITF have functional NLS. This renders *mi*-MITF incapable of transporting GABP• to the nucleus. Even more – *mi*-MITF can bind to and mislocalize also GABP•, regardless of the fact that it has its own NLS (Morii et al., 2001). The presence of a leucine zipper domain in GABP• led to the unproved speculation that MITF as a ZIP-protein may bind to this domain.

1.5.2. Interactions with GABP•
1.5.2.1. Host Cell Factor (HCF)
HCF-1 is another co-activator that physically binds GABP. Both, HCF-1 (Wilson et al., 1993) and its interaction with GABP were discovered in studies on regulation of the Immediate Early genes (IE) enhancer in Herpes Simplex Virus (HSV), where GABP is also involved. Instead of directly binding to DNA, HCF-1 connects by protein – protein interactions to •TIF – a viral transcription factor. It also binds to the TAD and leucine zipper domains of GABP•, but does not bind to GABP•• (shown by Yeast two-hybrid analysis). When not bound to GABP, the HCF-1 binding domain functions as a repressor, while after binding to GABP•, the complex is able to stimulate weak transcription (Vogel and Kristie, 2000).

1.5.2.2. YY1 and E4TF1 Associated Factor 1 (YEAF1) and YY1 associated factor (YAF-2)
YEAF1 and YAF-2 were both proven by yeast two-hybrid screens to bind GABP•. Similar to HCF-1, they do not bind GABP•. YEAF is attaching to a region in GABP•, situated in the middle of the molecule, which is common for all GABP• isoforms (Sawa et al., 2002). The domain in YEAF1, responsible for the connection is identified by deletion analysis and is called conserved region 2 (CR2). This is a zinc finger motif, homologous to YAF-2, which determines their similar way of binding. Their function however, is completely

Introduction

opposite to one another - YAF-2 is a co-activator and YEAF – repressor of the GABP dependent transcription (Sawa et al., 2002).

Furthermore, GABP• may form a ternary complex together with YEAF and YY1, as was revealed by yeast three-hybrid analysis. Most interestingly, in liver cells GABP and YY1 have opposite effects on the regulation of the HRS protein expression, where GABP activates the HRS promoter and YY1 is repressing it. Under differential conditions, YEAF1 may participate in these antagonistic effects, affecting one of the two transcription factors (Du et al., 1998).

1.5.2.3. E2F1

E2F1 is the only member from a group of transcription factors that can bind GABP• (Hauck et al., 2002). E2F-family factors are involved in apoptosis (Mundle and Saberwal, 2003) and in the G_1/S cell cycle transition. The family members can bind and temporarily inactivate members of the retinoblastoma gene family (pRb), thus effectively blocking entry into S phase. Upon phosphorylation, Rb releases the corresponding E2F factor and reverses its function.

In cardiac myocytes, E2F1 binds to GABP•$_1$-37, but not to GABP•$_1$-41 (Hauck et al., 2002). The authors observed also that a mutant GABP•$_1$-41, lacking its C-terminus is also able to interact with E2F1. This led to the suggestion that the C-terminal homodimerization domain (present in GABP•$_1$-41, but absent from GABP•$_1$-37) is interfering with and abolishing the connection with E2F1.

Other studies describe the functional effects of GABP•–E2F cooperation. While the E2F binding site alone has little effect on activating the Thymidylate Synthase promoter, removal of GABP and/or SP1 sites greatly diminishes the activity (Geng and Johnson, 1993 and Rudge and Johnson, 2002). The same three factors are activating the transcription of the catalytic 180-kDa subunit (Izumi et al., 2000) and the 54-kDa subunit (Nishikawa et al., 2001) of the mouse DNA polymerase •.

Finally, GABP• can also repress the pro-apoptotic function of E2F1, decreasing the levels of its target gene transcripts (as caspases -3 and -7), but by mechanisms different from inhibition of its transactivation capacity (Hauck et al., 2002).

1.6. The Aim and Experimental Design of the Project

GABP is a ubiquitous *Ets*-family transcription factor specified by very different expression levels in various cell types. Potential GABP target genes regulate such basic cellular processes as cellular energy metabolism, protein synthesis and cell cycle control. All these processes are tightly interconnected, both under normal and abnormal development, and together determine balanced growth properties of the cell.

The aim of our investigation was to determine the overall impact of GABP·•• expression levels on the growth properties of mouse fibroblast cells *in vitro*, using RNA-interference mediated down-regulation and both constitutive and conditional overexpression of GABP subunits.

2. RESULTS

The exhibited data represent outcomes from at least three independent experiments.

2.1. The *Gabp•* Gene Expression Effectively Silenced by RNAi Techniques

To down-regulate the GABP transcriptional activity we decided to silence GABP• expression using RNA-interference. Although in this case the protein of interest is not eliminated completely, residual amount is usually decreased to a degree, at which the function is lost or significantly impaired. In the particular case with GABP• knock-down, complete elimination of the protein is not desired. The search of GABP• mRNA sequence for the best potential siRNA target sites was performed using Ambion's "siRNA Target Finder and Design Tool" (http://www.ambion.com/techlib/misc/siRNA_finder.html). Within the GABP• coding region 96 potential targets were identified from which 7 candidates were chosen, on the basis of the listed criteria (see Appendix) and sequence homology between the human and mouse GABP• cDNA sequences (numbers 5, 7, 8, 10, 11, 18 and 25). Complementary synthetic oligonucleotides representing selected targets were hybridized and cloned into the siRNA vector pSUPER, which expresses siRNA hairpin structures under control of Histone 1 (H1) promoter (Brummelkamp et al., 2002 - Science) and several independent clones for each siRNA target were verified by sequencing.

These vectors were screened for their ability to down-regulate the expression of GABP• in U2OS cells. The screening was performed by transiently co-transfecting the cells with the siRNA vector together with pMSCVpuro which carries puromycin resistance gene. Successfully transfected cells were selected with puromycin. The whole cell protein extracts from transfected (survived) cells were compared on Western blot with the extract from cells, transfected with the empty pSUPER vector and pMSCVpuro only. The screening identified two siRNA-expression vectors, siGA7 and siGA10 (see Appendix), which strongly downregulated GABP• •expression (Fig. 2.1.-A).

In order to accomplish more efficient delivery of short interfering RNAs and to achieve stable down-regulation of GABP• the entire pSUPER expression cassette from the siGA10 knockdown vector was cloned into pMSCVpuro retroviral vector - pRetroSuper-GABP• 10 (pRS10) (Brummelkamp et al., 2002 - Cancer Cell). To enable detection of the transfected cells via GFP-fluorescence, EGFP-containing versions of retroviral vectors were created (pMSCVpuroG and pRS10G respectively). (See Appendix for clone charts.)

The pRS10 construct was tested for its ability to down-regulate the expression of GABP• in U2OS cells in transient and stable transfections. pMSCVpuroG and pRS10G constructs were used for production of viral particles for further infection of other cells of interest. Transient transfection/selection of U2OS cells with pRS10 resulted in a decrease of GABP• expression, yet to a lower extent compared to siGA10 and siGA7 (Fig. 2.1. B).

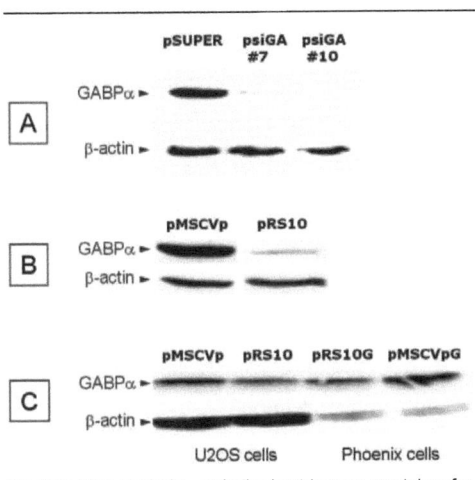

Fig. 2.1: Downregulation of GABP• by siRNA.

A, Substantial transient downregulation of GABP• expression in U2OS cells (85-fold for #7 and over 100-fold for #10) Transient transfection performed with pSUPER (empty vector) and the siRNA test-constructs (psiGA#7 and psiGA#10). B, GABP• downregulation in U2OS cells attained by transient transfection with the viral construct pRS10 (inserted as a plasmid). pMSCVp (empty vector) shows the endogenous GABP• level. The best dilution for the pRS10 vector reduces the GABP• amounts tenfold. C, Insufficient GABP• downregulation by pRS10 and pRS10G in stably transfected cells: 1,3 fold in U2OS cells and 1,15 fold in Phoenix cells after virus production. The results are confirmed for consistency by multiple (more than three) Western blots analysis – only the best images are taken for compiling the figure. Empty vectors – pSUPER, pMSCVp and pMSCVpG, •-actin serves as a loading control. The strength of the signal is determined by densitometry.

From the transiently transfected and selected U2OS cell populations polyclonal cell cultures were derived after prolonged (more than three weeks) selection with puromycin. These stably transfected cells were tested for expression of GABP• in Western blot. In contrast with transiently transfected cells, the stably transfected cells showed much weaker down-regulation of GABP• expression (Fig. 2.1 C). This indicated that there might be a problem to create a proliferating culture of cells with strong downregulation of GABP• expression (Yang et al., 2007).

The earlier created constructs pMSCVpuroG and pRS10G, bearing an additional EGFP marker gene, were used to transfect the Phoenix packaging cell line and to produce replication-deficient retroviral particles for further transduction of other cells of interest. The virus-producing Phoenix cells (>90% transfected cells) were also checked for GABP• protein levels but no substantial down-regulation caused by pRS10 and pRS10G was detected (Fig. 2.1-C).

Results

Taking into account that major down-regulation of GABP• is possible only in transient manner and significant decrease in the amount of this protein is not tolerated by the cells for long time we further chose to increase the cellular GABP content by additional expression of exogenous GABP.

2.2. Constitutive Expression of Exogenous GABP• Accelerates Proliferation of NIH-3T3 Cells

The GABP• subunit is crucial for the translocation of the whole GABP• •• complex into the nucleus. Therefore, the relative amount of GABP• in the cell is important for complete nuclear localization of the both subunits of the complex. Hence exogenous GABP• was additionally expressed in NIH-3T3 cells (see Fig. 2.2) to test the possibility if the increased nuclear amount of this subunit would influence cell proliferation speed.

To construct vector for stable expression of GABP•• entire cDNA was synthesized with PCR on the template of pRSV-GABP• using Pwo polymerase with proofreading activity and primers, specified in "Materials and Methods". This cDNA fragment was cloned into the pMSCVpuroG vector, described above (2.1.). Protein coding sequence of resulting vector, pMSCVpuroG-GABP••was completely sequenced.

2.2.1. Transgenic NIH-3T3 Cells Are Stably Overexpressing Exogenous GABP•

Several studies implicated GABP in the cell cycle regulation in NIH-3T3 cells by influencing the expression of some of critical cell-cycle regulators, such as Skp2 and Aurora A (Imaki et al., 2003 and Tanaka et al., 2002).

NIH-3T3 cells were transfected with pMSCVpuroG-GABP• and the parental pMSCVpuroG vector as a control; they were selected with high concentration of puromycin (3 µg/ml) for one week. Surviving cell populations were amplified and examined in Western blots. In sub-confluent cultures from NIH-3T3-GABP• cells, the observed exogenous expression of GABP• was around 10-fold higher then endogenous levels. A strong decrease of endogenous GABP• expression was observed in confluent cultures of control cells. Interestingly, a similar decrease was observed in confluent NIH-3T3-GABP• cells, where majority of detected GABP• is expressed from transgene (For possible reasons see chapter "Discussion".). No increase in the expression level of GABP• subunit was detected under these experimental settings. (Fig. 2.2)

Results

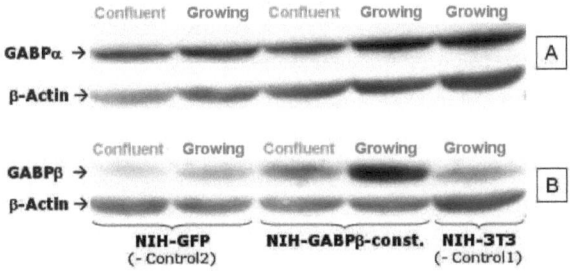

Fig. 2.2: Constitutive expression of exogenous GABP• in NIH-3T3 cells.
Western blots from polyclonal cell cultures of stably transfected NIH-3T3 cells with the vectors pMSCVpG (Control 2) and pMSCVpG-GABP• (NIH-3T3-GABP•-const.). The stably transfected cells are constantly grown in presence of selective agent. The non-transfected wild type NIH-3T3 cells are used as empty control (Control 1); loading control is •-actin. Visualisation of: **A**, GABP• expression and **B**, GABP• expression in growing and confluent cell cultures. Strong expression of exogenous GABP• is observed only in growing cultures and GABP• expression is not affected by the additional expression of GABP•. The overexpression of both GABP• and GABP• is observed in multiple occasions on Western blots, the current Western blot is specifically designed for visualization purposes.

2.2.2. The Overexpressed GABP• Protein Demonstrates Effective Nuclear Targeting

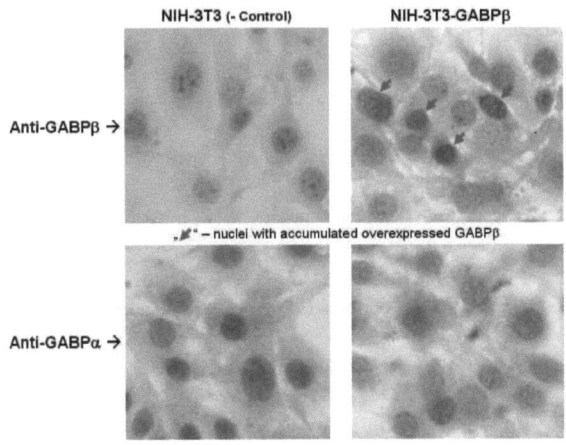

Fig. 2.3: Predominant nuclear localization of the constitutively overexpressed, exogenous GABP•.
A polyclonal cell culture of stably transfected NIH-3T3 cells with the vector pMSCVpG-GABP• is visualized by cytostaining. The antibody against GABP•₁ shows a very distinguished accumulation of this protein in some of the nuclei. Arrows indicate nuclei with strong GABP• accumulation. This is one out of five cytostaining experiments, performed in various combinations and at different times.

The sub-cellular localization of GABP subunits was determined by cytostaining of control and exogenously GABP•-expressing transgenic cell cultures with antibodies directed against GABP•. and GABP•. NIH-3T3-GABP• cells exhibit visibly increased amounts of nuclear GABP• •Fig. 2.3). Different levels of GABP• staining between individual nuclei are observed in the figure, due to the heterogeneity of the culture. The stained cells are derived from many transfection events and consequently are possessing different properties. Any GABP• overexpression, correlating with the GABP• overexpression in this experiment was not observed. Although the growth properties of the derived culture were influenced (see below), additional recruitment of GABP• from the cytosol into the nucleus could not be detected.

2.2.3. Constitutive GABP• Overexpression Accelerates Cell Proliferation

Involvement of GABP in cell cycle regulation (Imaki et al., 2003 and Tanaka et al., 2002) and our observation that expression of endogenous GABP• is decreased in confluent NIH-3T3 cultures suggested a link between GABP• expression level and culture growth speed.

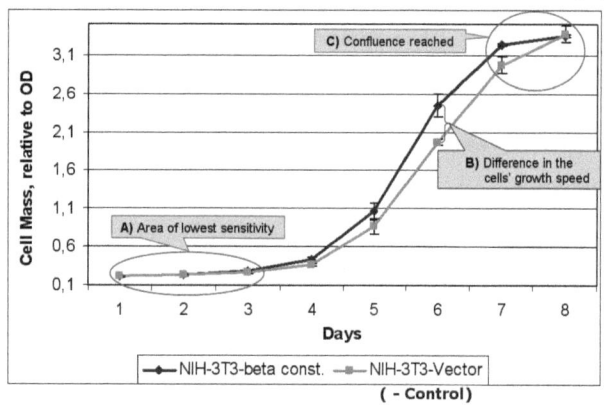

Fig. 2.4: Proliferation rate of NIH-3T3 cells constitutively overexpressing GABP•. Increase in the Proliferation rate of GABP• overexpressing cells was detected using Crystal violet assay. "NIH-3T3-GABP•-const." cells are stably transfected NIH-3T3 cells with pMSCVpG-GABP•. For negative control NIH-3T3 cells were used stably transfected with the vector pMSCVpG (empty vector). **A,** In the early course of the experiment (first three days) no difference in the signals could be detected. **B,** With the highest sensitivity of the method reached at day six, the difference between the signals of the "Control" and the "Sample" was the best pronounced. **C,** Both cell cultures became confluent at the latest time points of the experiment.

The scheme depicts the most representative data from three independent experiments (totally five performed). Each point is comprised from the average value of three repeats. Error bars represent Standard Deviation between the repeats. Statistical significance is calculated as Paired T-test, two-tailed distribution. The results are significant as the p-value<0,01 (P=0,004029).

Crystal Violet staining was used to determine the cell mass of different cell types at defined time points. As in the course of the experiment no change in the size of single cells was observed, a change in the final signal could only be explained by differences in the total cell number. Equal cell numbers were plated and grown under identical conditions for defined periods of time. NIH-3T3-GABP• grew around 20% faster than control NIH-3T3-Vector cells (Fig. 2.4).

2.3. Newly Created Vectors Direct Simultaneous Doxycycline-Regulated Expression of GABP• and GABP•

Constitutive overexpression of GABP• resulted in a slightly accelerated cell proliferation speed. However, GABP is heterotetrameric transcription factor and in order to substantially increase transcriptional activity, both GABP• and GABP• subunits should be expressed simultaneously in comparable levels. In addition, conditional elevated expression would be advantageous to minimize adaptation processes in cells and to provide better negative controls. Therefore, for achieving these goals we chose to use the *Tetracycline*-regulated system (see Fig. 2.5).

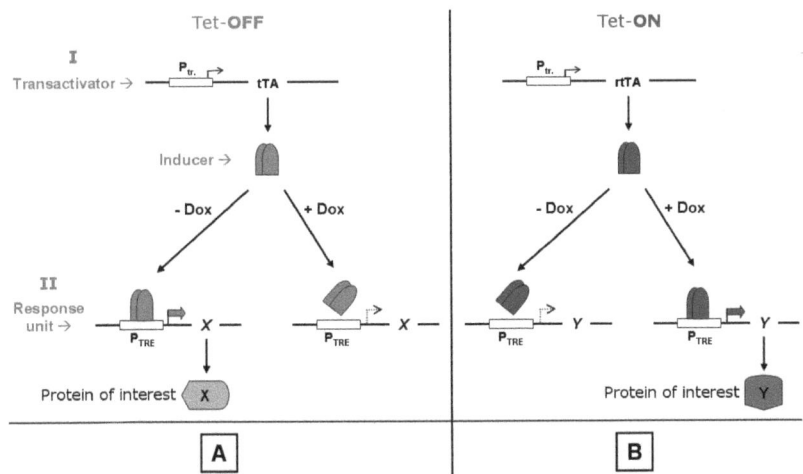

Fig. 2.5: Inducible regulation of exogenous gene's expression by *Tetracycline* Responsive Element (TRE) – mechanism of action: The system consists of two elements: **I** - the Transactivator (small protein regulatory element, which is constitutively expressed in the cells in minor amounts) and **II** – the Response unit (the gene of interest). They interact with each-other with the help of a so-called Inducer molecule (*Tetracycline* or *doxycycline*). The transactivators can be of two types: **A**, Direct – tTA (binds to the Response unit and activates the expression in the *absence* of Inducer, called also *tetracycline*-off or Tet-OFF) and **B**, Reversed – rtTA (binds to the Response unit and activates the expression in *presence* of Inducer, called also *tetracycline*-on or Tet-ON). **X, Y** – Proteins of interest. The figure is adapted from: http://www.tetsystems.com/science-technology/scientific-figures/

Results

On its basis, several *Tetracycline*-Inducible constructs were created, able to conditionally express GABP• and/or GABP• protein (For clone charts see the Appendix).

By cloning into the plasmid pTRE-6xHN were created two other constructs: pTRE-GABP• (expressing GABP• under the regulation of *Tetracycline* Responsive Element (TRE) and pTRE-GABP• (analogous, expressing GABP••. Using these two plasmids as source of DNA inserts and the earlier created pMSCVpuro-• plasmid (on the basis of pMSCVpuro) as a vector, were constructed two additional plasmids: pMSCVpuro-• -TRE-GABP• (self-inactivating retroviral vector, expressing GABP• under the regulation of TRE) and pMSCVpuro-• -TRE-GABP• •• (self-inactivating retroviral vector, expressing both GABP• and GABP• under the regulation of TRE).

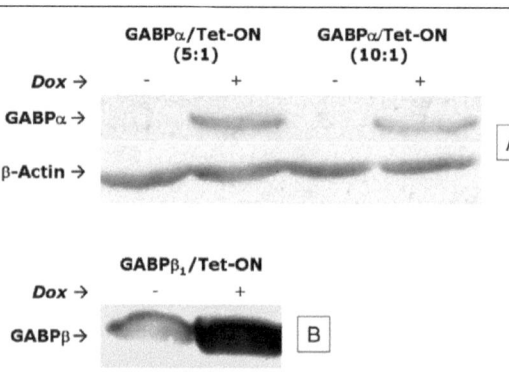

Fig. 2.6: Successful conditional expression of GABP• and GABP• by vectors encoding either GABP• or GABP• .

Phoenix cells were transiently co-transfected with one of the parental plasmids pTRE-GABP• or pTRE-GABP• together with pTet-ON, induced with *Doxycycline* for 24h and examined on Western blot. **A**, The GABP• and pTet-ON vectors were mixed in ratios 5:1 and 10:1. Antibody used for the Western blot – anti GABP• • loading control – •-actin. **B**, The GABP• expression is visualized with antibody against GABP•.. The figure is compiled from multiple Western blots; representative samples of qualitatively similar Western blots are shown.

In order to prove the *Doxycycline* inducibility of the created constructs were performed series of transient transfections and Western blot analysis. Phoenix cells were transiently co-transfected with the vector of interest and plasmid, constitutively expressing the Transactivator. After 24 hours of induction with *Doxycycline*, whole cell extract was prepared and probed on Western blot with antibodies directed against GABP• and GABP• . Selected constructs efficiently directed *Doxycycline*-inducible expression of GABP• and GABP• (Fig. 2.6). Similar results were observed with the retroviral vectors, expressing only one or both subunits (Fig. 2.7).

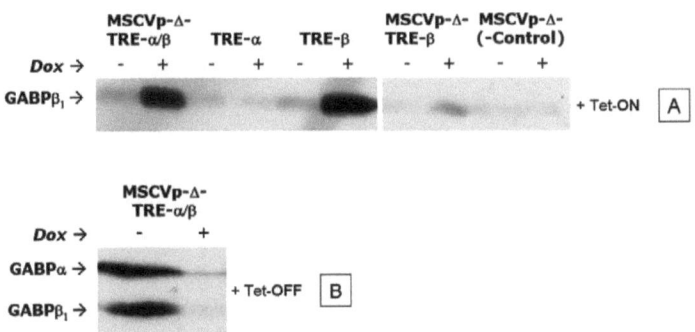

Fig. 2.7: **Successful conditional expression of GABP single and double constructs.** Phoenix cells were transiently co-transfected with one of the GABP-expressing plasmids together with pTet-ON or pTet-OFF, induced with *Doxycycline* for 24h and examined on Western blot. **A,** Various GABP-expressing constructs, visualized with anti-GABP•₁ antibody. All samples were co-transfected with pTet-ON transactivator. p•MSCVp is the empty vector (negative control). **B,** Viral construct, conditionally expressing GABP• plus GABP• (pMSCVp-•-TRE-GABP•••). This sample is co-transfected with pTet-OFF transactivator.
The figure is compiled from multiple Western blots – the data are confirmed through a series of successive blots.

2.4. Stably Transfected NIH-3T3 Cells Conditionally Express High Amounts of GABP• and GABP•

Stably transfected cells, bearing the transactivator of the TET-system and vectors necessary for the conditional expression of the both GABP subunits under TRE-regulation had to be created. Those vectors are the following: pTetOn-S2, which carries an improved variant of the reverse *Tetracycline* transactivator (rtTA-S2) and also the plasmids pMSCVpuro-•-TRE-GABP• and pMSCVpuro-•-TRE-GABP••• described above, bearing the conditionally regulated genes of interest.

The creation of the double-transfected cells was executed on two stages, by two consecutive transfection events, each followed by selection for the stably transfected cells, single cell cloning and selection for the best performing clones at every stage.

Results

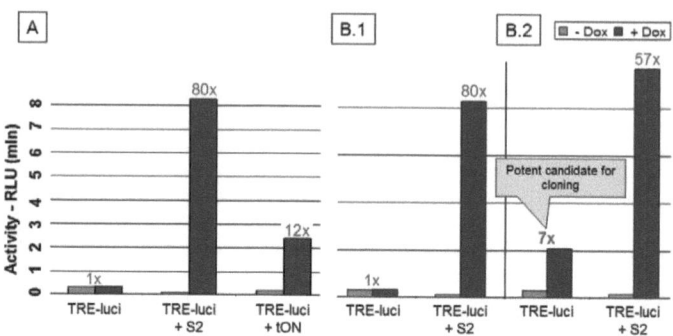

Fig. 2.8: Efficiency of two different types of transactivators. All cell types were transiently co-transfected with the corresponding plasmids. Each transfection is performed in triplets and the samples were averaged by pooling before measurement. Half of the cells were induced with *Doxycycline* for 24h and examined by Luciferase assay. For negative controls, transfections with the pTRE-luci were done. The fold induction is shown in red numbers above the columns. **A**, Evaluation of the transactivators' activity. NIH-3T3 cells were transiently co-transfected with pTRE-luci and pTet-ON-S2 or pTet-ON, where Tet-ON-S2 shows almost seven-fold higher inducibility. **B**, Determining the transactivating potential of stably transfected NIH-3T3-S2 cells. Two type cells were transiently co-transfected with the constructs pTRE-luci and pTet-ON-S2. **B.1**, NIH-3T3 wild type cells were used as control. **B.2**, The transfected NIH-3T3-S2 non-cloned cell pool culture shows seven-fold inducibility and is therefore a potent candidate for cloning.
Similar results were obtained using HeLa cells as an independent cell source for confirmation purposes only. The graph is compiled from representative data of several experiments.

The existence of two forms of TET transactivators (for references see http://www.clontech.com) raised the question of determining the more appropriate one for further experimental procedures. Check-up experiments were set to confirm the better suitability of one of them for the cell/construct system used.

NIH-3T3 cells were transiently co-transfected with the Firefly luciferase reporter construct pTRE-luci and one of the two transactivators pTetON or pTetON-S2. Afterwards half of each sample was induced with *Doxycycline* for 24 hours and the transfected cells were examined in luciferase assays. The intensity of the emitted light was measured in Relative Light Units (RLU). As a negative control, only pTRE-luci transfected cells were used (Fig. 2.8-A). The choice fell on S2 transactivator as it showed seven fold higher inducibility compared to TetON.

2.4.1. Transgenic NIH-3T3-S2 Cells Constitutively Express S2 Transactivator

To create the NIH-3T3-GABP• •• cells, wild type NIH-3T3 cells were co-transfected with S2 transactivator expression vector and pMSCVneo (ratio 10:1). Transfected cells were selected with neomycin for one week and expanded under continuous presence of neomycin. To examine the transactivation potential these polyclonal cultures were tested in transient transfection assays (Fig. 2.8-B.2), using luciferase reporter construct under control of *tetracycline* responsive elements (TREs). As a positive control co-transfection with S2 transactivator expression construct was used. We observed a seven fold increase in the luciferase activity when *Doxycycline* was added to the pooled S2 transgenic cells indicating proper operation of Tet-system at least in a subpopulation of these cells. These observations suggested that individual clones with relatively high induction ratio might be isolated from this population.

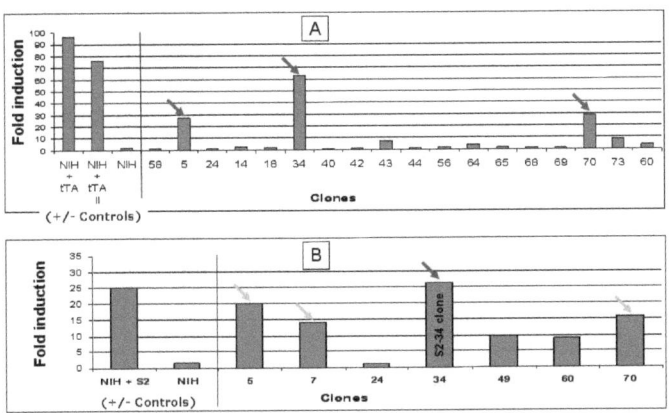

Fig. 2.9: Screening for transactivator's expression amongst Tet-ON-S2 cells results in detection of highly inducible clones. NIH-3T3 cells were stably transfected with pTet-ON-S2 and single-cell cloned. Many clones were obtained (less than hundred). All of the clones were transiently transfected with pTRE-luci, induced with *Doxycycline* for 24h and examined by Luciferase assay. For positive controls NIH-3T3 cells were used co-transfected with reporter and transactivator and for negative controls - NIH-3T3 cells transfected with reporter only. The inducibility is shown as the quotient of the induced and non-induced sample values from each clone. Arrows indicate the best performing clones. **A,** Example of screening for S2 transactivator activity in single-cell cloned stable transfected NIH-3T3 cells is shown. **B,** Confirmation for consistency of the S2 transactivator activity by repeated transfection of the best clones. Clone S2-34 showed the highest inducibility and was used for further experiments.
The graph is compiled, using results from four independent transfections, each with 16 to 21 variants. Each transfection was performed in three separate wells and the results were averaged by pooling the samples before measurement.

In order to obtain individual clones, S2-transactivator expressing population of NIH-3T3 cells was plated in flat-bottom 96-well plates at density 0,3 – 0,5 cells per well and cultivated in the presence of neomycin until visible appearance of colonies. Wells with single-cell colonies were expanded and screened for *doxycycline*-inducible activity of luciferase reporter, as described above. The clones with the highest inducibility results were selected for the second round of identical screening (Fig. 2.9-A, -B). Thus ten

Results

clones (out of nearly hundred examined) were found to have a high rate of inducibility, and clone #34 was selected for further work (Fig. 2.9-B).

2.4.2. Double Transfected Cells Overexpress GABP• ••

Clone NIH-3T3-S2 #34 was used for the introduction of the constructs, conditionally expressing GABP• and GABP•, or GABP• only. (pMSCVpuro-• -TRE-GABP• •• and pMSCVpuro-• -TRE-GABP•).

The cells were transfected and selected with puromycin for three weeks, followed by the isolation and expansion of single-cell derived clones as described above. In addition, inducible overexpression of GABP• •• was verified using Western blots.

Numerous clones expressing GABP• (with full name NIH-3T3-S2-34-GABP•, but from now on called NIH-3T3-GABP• or GABP• cells) or GABP• • • (with full name NIH-3T3-S2-34-GABP• ••, but from now on called NIH-3T3-GABP• •• GABP• ••• or shortly • •• cells) to different extents were found and several were used in the later studies (Fig. 2.10).

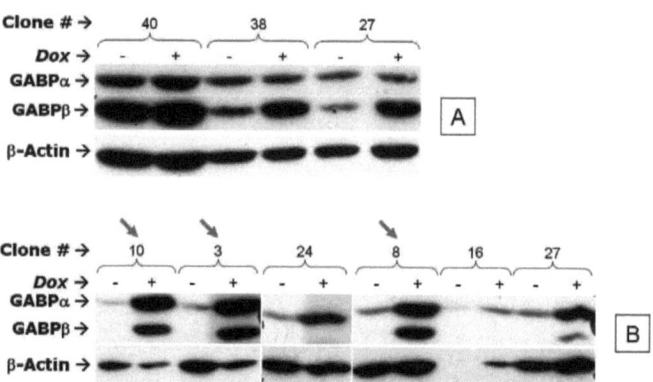

Fig. 2.10: Example of screening for inducible expression of GABP• •• and GABP• in individual NIH-3T3-S2 clones. NIH-3T3-S2 cells were transfected with pMSCVpuro-• -TRE-GABP• and pMSCVpuro-• -TRE-GABP• •• and single-cell cloned. Half of each sample was induced with *Doxycycline* for 24h and both parts were examined on Western blots. •-actin was used for loading control. Arrows indicate the best performing clones. **A,** Part of the GABP• clones are shown – the NIH-3T3-S2-34 cells are transfected with pMSCVp-• -TRE-GABP•. **B,** Part of the GABP• •• clones are shown – the NIH-3T3-S2-34 cells transfected with pMSCVp-• -TRE-GABP• ••.

For each transfection many more clones were examined (typically less than hundred). Images from several independent Western blots are used for this figure.

2.5. The Inducible GABP• •• Proteins Demonstrate High Levels of Overexpression and Effective Nuclear Targeting

To characterize sub-cellular localization of the exogenously expressed GABP• and GABP• proteins the cells from clone • ••• 3 (with the highest level of inducible expression) and NIH-3T3-S2-34 cells (the parent cell culture, serving as control for endogenous GABP levels and side-effects of *Doxycycline* treatment) were cultivated in the presence or absence of *doxycycline*, fixed and stained with the antibodies directed against GABP• or GABP•.

With both antibodies used the clone • ••• 3 displayed a clear and strong increase in nuclear appearance of both GABP• and GABP• after *Doxycycline* treatment, while parental S2-34 cells showed no difference in expression and sub-cellular localization of the GABP subunits (Fig. 2.11)

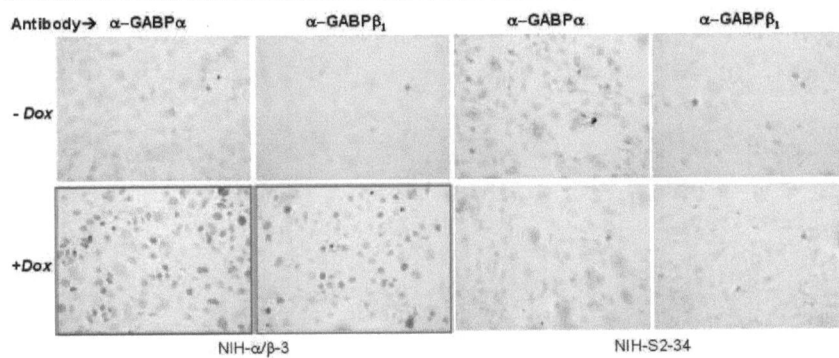

Fig. 2.11: Nuclear localization of the conditionally overexpressed GABP• and GABP•. The intracellular localization of GABP• and GABP• is visualized by cytostaining on growing cultures from the cell clone NIH-3T3-GABP• ••-3 compared to the parental NIH-S2-34 cells. Both cell types are stained with anti-GABP• and anti-GABP•$_1$ antibodies and half of them are induced with *Doxycycline* for 36h. No DNA staining is performed on the shown samples to avoid masking of the protein signal. Predominantly nuclear localization of the both subunits is observed and in case of overexpression the signal is many times stronger.
In total five cytostains were performed, at different times and in various combinations of staining agents/antibody. Pictures in magnifications 10x and 40x were taken, from which only the latter were used in the figures. The results are consistent throughout the different experiments. Zoomed-in version can be seen at Fig. 2.25.

It should be noted that during routine checks of the accumulated data a small divergence in the sequence of one of the GABP subunits was observed. A point mutation at the very end of the coding sequence of GABP• in the plasmid pMSCVpuro-•-TRE-GABP••• resulted in a transformation of the stop codon (**TGA**) into the Gly codon (**GGA**) and extension of coding sequence by six extra amino acid residues. This C-terminal extension of GABP• does not appear to influence any of the protein's functional properties examined [complete expression, nuclear localization, binding to GABP• and other GABP• subunits, DNA-binding of the complex (see below) and its activity]. In all further experiments the possibility for such an influence was taken into consideration and rejected.

2.6. The Exogenous GABP• •• Subunits Possess Strong DNA-Binding Activity *in Vitro*

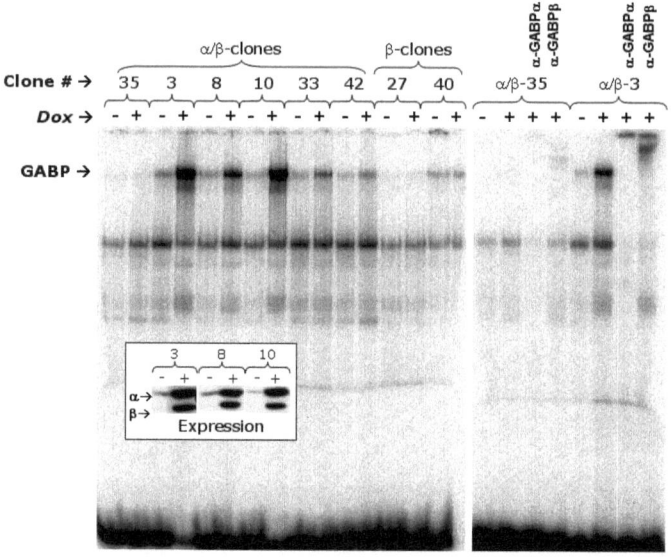

Fig. 2.12: The exogenous GABP• •• subunits strongly bind to their corresponding DNA motif. Various cell clones were induced with *Doxycycline* for 36h and nuclear extracts were prepared and examined in Electrophoretic Mobility Shift Assays (EMSAs). The cell clones used are as follows: A clone, conditionally overexpressing GABP• only – •-27, a clone constitutively overexpressing GABP• only – •-40, five clones expressing different amounts of GABP• and •• together – • ••-clones numbers 3, 8, 10, 33 and 42 and the clone • ••-35, which is transfected with the same constructs as the rest of the GABP• •• clones, but does not show any overexpression and is used as a negative control. The insert shows Western blot signals from the protein expression of the three most active clones. Anti-GABP• and anti-GABP•₁ antibodies are used to induce supershifting and to prove the specificity of the binding. Probe - DSE_{TSC2}: 5'-TCCGCTA**CCGGAAG**TGCGGGTCGCG**CTTCCGG**CGGCGT-3'

The data, shown in this figure are confirmed by several independent EMSAs.

Results

To characterize DNA binding activity of exogenously expressed GABP••• several clones with different levels of inducibility were chosen. The cells were cultivated for 36 hours in the presence or absence of *Doxycycline* and then harvested, nuclear extracts were prepared and used in Electrophoretic Mobility Shift Assays (EMSA) with a tandem GABP•••• binding site as a probe (Andris Avots, unpublished). Antibodies directed against GABP• and/or GABP• were included in selected binding reactions to prove specificity of observed complexes.

Strong increase of specific GABP•••• binding activity was observed in nuclear extracts from *Doxycycline* treated cells. Importantly, the level of DNA binding activity correlated well with the level of overall GABP•••• expression detected in Western blots (Fig. 2.12).

Doxycycline-inducible increase of GABP•••• binding activity was neither detected in nuclear extracts from the cells conditionally overexpressing GABP• only (•-27), nor in those used as controls (•••-35 and •-40). This indicates that at least in NIH-3T3 cells there is no substantial extra cytosolic fraction of GABP• which could be mobilized into the nucleus by increased expression of •-subunit.

2.7. Expression of Excessive Amounts GABP•••• Influences the Proliferation Speed of NIH-3T3 Cells

Earlier data suggest involvement of GABP•••• complex in the regulation of cell proliferation via regulation of cell cycle progression (Imaki et al., 2003 and Tanaka et al., 2002). Therefore we investigated how the elevated GABP expression would influence cell proliferation speed. Crystal Violet staining of proliferating cell cultures in the presence and absence of *Doxycycline* was used to determine the relative change in the cell mass. This accurately correlates with the number of cells if the cell size is not altered.

2.7.1. Conditionally Increasing the GABP•••• Expression Does Not Influence the Size of NIH-3T3 Cells

Flow cytometry analyses were performed to determine if the cell size is altered by a rise in GABP expression. Four transgenic NIH-3T3 clones conditionally expressing GABP•••• at different levels were cultivated for 72h in the presence or absence of *Doxycycline*. In NIH-3T3 cells pronounced accumulation of the expressed GABP subunits requires less than 24 hours (steadily increasing with the time up to 3 days - Fig. 2.27) and cell's

Results

doubling time is app. 20 hours, therefore incubation of 72h ensures about three cell divisions in the presence of increasingly higher GABP levels. This time is sufficient for potential changes in the cell size to occur, if any.

The different samples were compared in the mean values of their Forward scatter (FSc), characterizing the approximate cell size. Then the percentage differences between the induced and non-induced samples were calculated as well as their Standard deviation (St. Dev.). Thus very small differences in the cell sizes were found, the St. Dev. being only 2,7%. Even less influence was found when Orthogonal/Side scatter (SSc) values, which depict complexity or granularity of the cells, were compared (St. Dev. 2,6%). Therefore *Doxycycline* treatment and increased expression of GABP• •• did not change either average size or granularity of NIH-3T3 cells (Fig. 2.13) and Crystal Violet assays could be employed to characterize the proliferation kinetics of these cells.

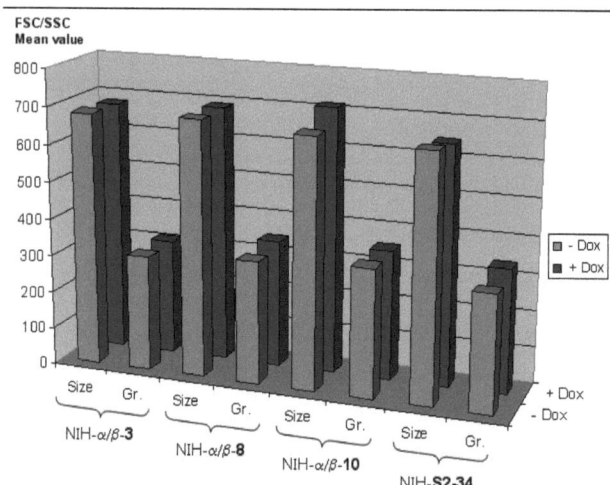

Fig. 2.13: The size and granularity of NIH-3T3 cells does not change during exogenous GABP• ••- expression. The cell clones NIH-3T3-GABP• ••-3, -8 and -10 are induced with *Doxycycline* for 36h and examined by flow cytometry, for negative control the parental cells NIH-3T3-S2-34 are used. The samples are compared by their mean values, gated on the living cell population. One representative experiment is shown. When harvested, all cells were in exponential growth phase with the same seeding densities.

2.7.2. Conditionally Elevated GABP• •• Expression Results in a Reversible Reduction of Proliferation Speed of NIH-3T3 Cells

To determine the effect of additional amounts of GABP• •• on growth properties of NIH-3T3 cell cultures individual clones with different levels of inducible GABP• •• expression were plated at low densities and cultivated in the presence or absence of *Doxycycline*. Cell mass was determined at the indicated time-points (figures 2.14-2.16) using Crystal Violet assay (see the appendix for graphs from more Crystal Violet assays).

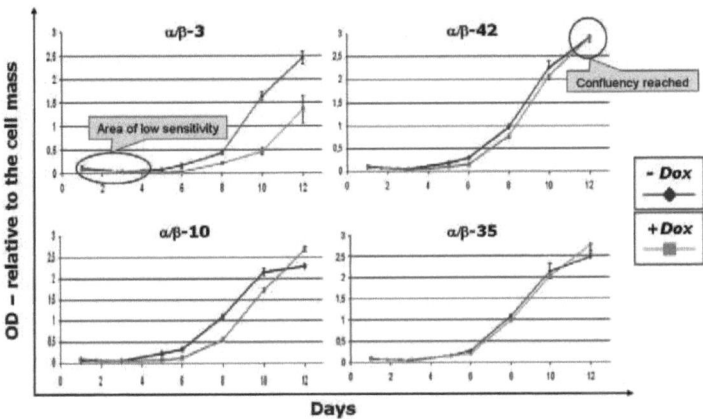

Fig. 2.14: Increased GABP•⁂ expression results in decrease in proliferation speed of NIH-3T3 cells. The cell clones NIH-3T3-GABP•⁂-3, -10 and -42 are grown at low densities (1250 cells/well, in 12 well dishes) for 12 days under differential conditions (in presence or absence of *Doxycycline*) and examined by Crystal Violet assays. For negative control the non-overexpressing clone •⁂-35 is used. In the various clones GABP•⁂ is expressed at different levels, ordered as follows: Clone •⁂-3 > •⁂-10 > •⁂-42 > •⁂-35. Note that higher level of inducible GABP•⁂ expression correlates with slower culture growth. The differences in the growth speed appear after the first four days of the experiment (area of low sensitivity). Most of the cell lines became confluent at the latest time points – after day 12.
The scheme depicts data from five independent experiments with partially overlapping conditions (seeding densities, days of incubation) and total of ten Crystal Violet assays. Qualitatively same result was obtained using different seeding densities. Each point is comprised from the common mean value of three separately measured samples. Error bars represent the Standard Deviation between triplicates. For the time point with best differing values statistical significance is calculated as Homoscedastic T-test (two-sample equal variance) with two-tailed distribution. For sample •⁂-3 the results are significant as the p-value<0,01 (P=0,004793). The same statistical significance applies also for the figures 2.15 to 2.17 plus Fig. 2.19. The scheme shows data from a single representative experiment.

We observed that increased level of GABP•⁂ expression substantially slowed down the proliferation of the cells (Fig. 2.14), and that this reduction strictly correlated with the total level of GABP•⁂ expression and the level of GABP•⁂ binding activity in cells (Fig. 2.12). At beginning of the assay (days 1-4) only subtle differences were detected. This was likely because of prolonged lag-phase due to low plating density of cells. Therefore experiments were repeated at higher plating density of the cells (Fig. 2.15). Clear differences were observed already at day 2, even before the expression level of exogenous GABP•⁂ reaches maximum (Fig. 2.27). At the end of the incubation period the fastest growing cultures were reaching confluence, leading to contact inhibition and terminating the normal cell division. When 18 fold higher starting cell densities were used, the confluence was reached sooner (day 6-7 instead of day 10-11) and therefore the divergence in the signals of induced and non-induced cultures was not so well pronounced (Fig. 2.15).

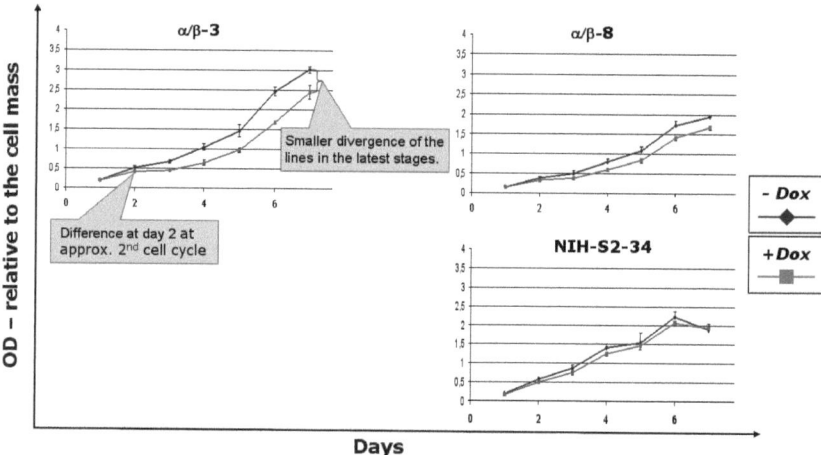

Fig. 2.15: Increasing the temporal sensitivity of the method by increase in the starting cell density. The cell clones NIH-3T3-GABP•••-3 and -8 are grown at 18 fold higher starting densities (22500 cells/well, in 12 well dishes) for 7 days under differential conditions (in presence or absence of *Doxycycline*) and examined by Crystal Violet assay. For negative control are used the parental cells NIH-3T3-S2-34. With higher cell densities the differences in the culture growth speed are visible already at day two, although the divergence between the induced and non-induced cells remains smaller at the final stages of the experiment due to the faster reaching of confluence.

The scheme depicts data from five independent experiments with partially overlapping conditions (seeding densities, days of incubation) and total of ten Crystal Violet assays. . Qualitatively same result was obtained using different seeding densities. Each point is comprised from the common mean value of three separately measured samples. The error bars represent the Standard Deviation between the repeats. The scheme shows data from a single representative experiment.

In the course of the experiment the difference between induced and non-induced cells was increasing, indicating that the culture growth speed was possibly time dependent. Therefore we compared the proliferation rate of NIH-3T3-GABP••• cell clones pre-incubated with *Doxycycline* for several days before the cells to be plated for proliferation experiment. As expected, induction of GABP••• expression for four days before the actual experiment resulted in an even stronger decrease of the culture growth speed (Fig. 2.16).

Results

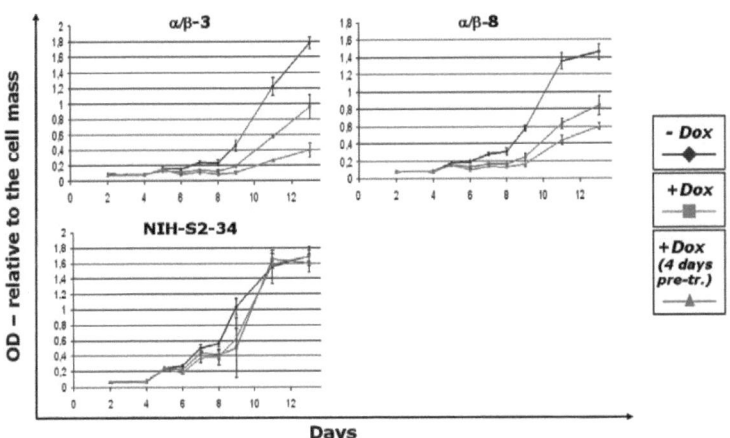

Fig. 2.16: Further decrease of the culture growth speed by prolonged exogenous GABP• •• **expression.** The cell clones NIH-3T3-GABP• •• -3 and -8 are grown at low densities (1250 cells/well, in 12 well dishes) for 13 days under differential conditions (in presence or absence of *Doxycycline*) and examined by Crystal Violet assay. For negative control the parental cells NIH-3T3-S2-34 were used. The cells, used for one of the samples were grown in presence of *Doxycycline* for four days prior the experiment. During the pre-treatment the cell cultures were maintained at normal growing densities, i.e. at about 50-70% confluence. Note that for maximum accumulation of exogenous GABP• •• only three days are required (Fig. 2.27).
The scheme depicts data from five independent experiments with partially overlapping conditions (seeding densities, days of incubation) and total of ten Crystal Violet assays. . Qualitatively same result was obtained using different seeding densities. Each point is comprised from the common mean value of three separately measured samples. The error bars represent the Standard Deviation between the repeats. The scheme shows data from a single representative experiment.

Aiming to check if the reduction of cell proliferation rate is reversible, parallel cultures of GABP• •• -3 and control cells were cultivated in the presence or absence of *Doxycycline* for four days, after which *Doxycycline* was withdrawn from one set of cultures. This resulted in a complete restore of culture growth speed reduction of • •• -3 cells, indicating reversibility of suppressive effects by GABP• •• (Fig. 2.17).

In conclusion, the expression of extra amounts of GABP leads to reversible suppression of the proliferation speed of NIH-3T3 cells under normal culturing conditions.

During the course of the Crystal Violet Assays formation of colony-like conglomerates (or clumps) of the growing cells was observed. This specific growth distribution was dependent on the level of GABP expression. In the presence of inducer the cell conglomerates were showing decreased number and size, when compared to controls. Decreased growth speed of the clones able to highly express GABP• •• was also noted even in the absence of inducer (Fig. 2.18).

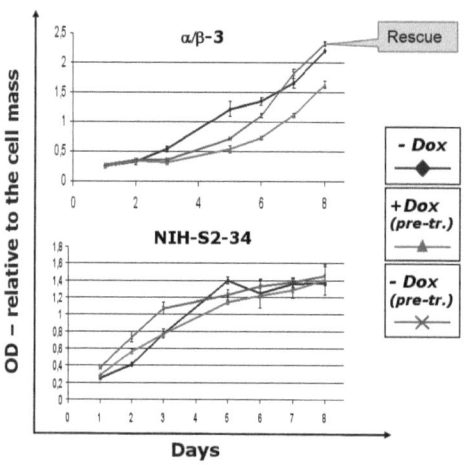

Fig. 2.17: Proliferation speed recovery, following withdrawal of the excessive GABP•••. The cell clone NIH-3T3-GABP•••-3 is grown at high densities (22400 cells/well, in 12 well dishes) for 8 days under differential conditions (in presence or absence of *doxycycline*) and examined by Crystal Violet assays. For negative control the parental cells NIH-3T3-S2-34 were used. Two of the samples were pre-treated with *Doxycycline* for four days prior the beginning of the experiment (labelled on the graph's legend with "(pre-tr.)"). At the beginning of the experiment the inducer was removed from one of the pre-treated samples (- Dox (pre-tr.)"). The growth speed of this sample starts to increase after three days and returns to its original values by day 7.

The scheme depicts data from five independent experiments with partially overlapping conditions (seeding densities, days of incubation) and total of ten Crystal Violet assays.
Qualitatively same result was obtained using different seeding densities. Each point is comprised from the common mean value of three separately measured samples. The error bars represent the Standard Deviation between the repeats. The scheme shows data from a single representative experiment.

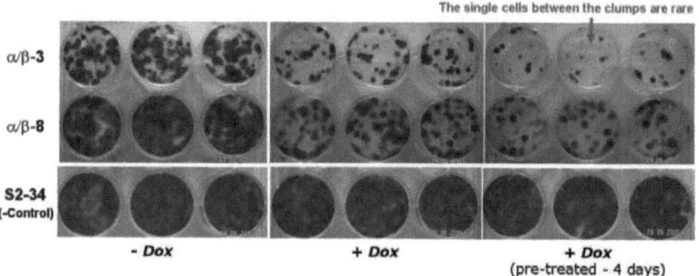

Fig. 2.18: Exogenous GABP••• expressing clones during growth form colony-like conglomerates/clumps. The cell clones NIH-3T3-GABP•••-3 and -8 were grown at equal (low) densities (1250 cells/well, in 12 well dishes) for 7 days in the presence or absence of *Doxycycline*. The cultures were then stained with Crystal Violet and photographed. For negative control the parental cells NIH-3T3-S2-34 were used. One of the samples was pre-treated with *Doxycycline* for four days prior the beginning of the experiment. The first panel (-Dox) shows the differences between the speed of the clone growth without induction.
The picture exemplifies a typical view of Crystal Violet assay plates after the assay is performed. The plates from all ten experiments have similar appearance.

As slightly higher basic GABP•••• expression was anticipated from some clones due to promoter leakage we can't rule out the possibility that the observed lower proliferation speed is caused by increased basic levels of non-induced GABP•••. As seen on the EMSA, clones •••• 3 and •••• 8 show higher basic binding activity and thus probably express more GABP even in the absence of *Doxycycline* treatment (Fig. 2.12).

To prove this assumption we have compared the basic growth speed of various clones conditionally expressing GABP•••, showing that clones •••-3 and •••-8 exhibit detectably lower growth speed even without induction (Fig. 2.19). As controls we used not only the parental cell line NIH-3T3-S2-34, but also the identically treated clones •••35 and •••42, which proves that the observed difference in the culture growth speed is not due to differences in the clones' history.

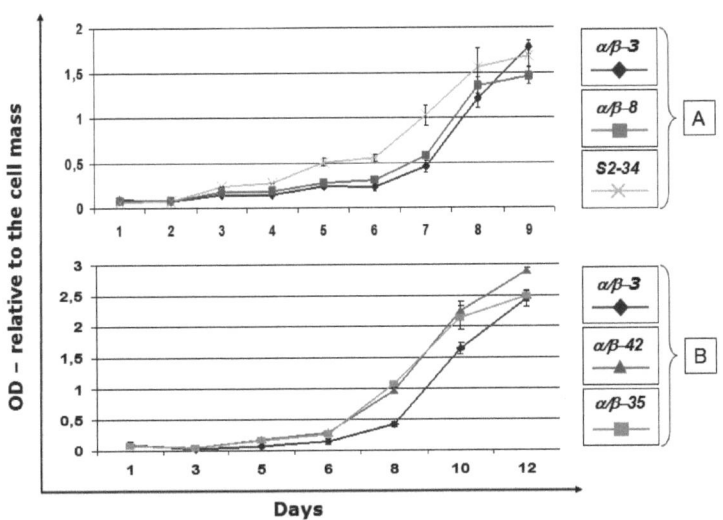

Fig. 2.19: The clone NIH-3T3-GABP•••-3, highly expressing GABP••• demonstrates decreased growth speed without induction. The indicated NIH-3T3-GABP••• cell clones were grown for the designated time in absence of *Doxycycline* and examined by Crystal Violet assays. **A,** The cell clones •••-3 and •••-8 were grown at high starting density (22500 cells/well in 12-well dishes) for 9 days. For negative control the parental cells S2-34 were used. **B,** The cell clones •••-3, and •••-42 were grown at low starting density (1250 cells/well, in 12-well dishes) for 12 days. For negative control the non-overexpressing clone •••-35 is used. The clones which are able to express higher amounts GABP••• exhibit slower culture growth even without induction of GABP••• expression (clones 3 and 8). Most of the cell lines became confluent at the latest time points – after days 8 and 10 respectively.

This figure is compiled from the data of several experiments with partially overlapping conditions (seeding densities, days of incubation). Total of ten Crystal Violet assays were accomplished. Each point is comprised from the common mean value of three separately measured samples. The error bars represent the Standard Deviation between the repeats.

This observation raised the question: what are the underlying molecular mechanisms? Was the cell proliferation slowed down, or the apoptosis was upregulated? To address this question, further experiments were conducted with the purpose to investigate possible changes in the cell cycle distribution or the rate of apoptosis in cultures expressing higher amounts GABP•••.

2.8. GABP•••• Overexpression Does Not Affect the Cell Cycle

Addressing the question if the decreased proliferation speed of cells expressing more GABP•••• is due to cell cycle regulation, we performed cell cycle distribution and cyclin expression analyses.

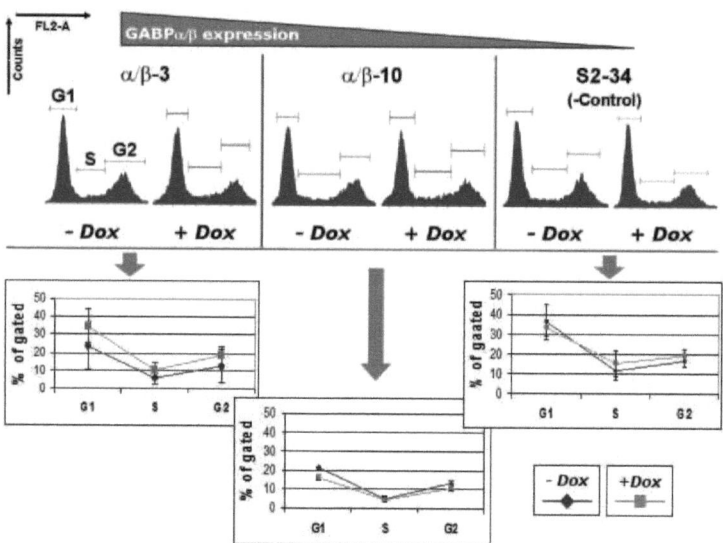

Fig. 2.20: Conditionally increasing the GABP•••• levels does not result in changes in cell cycle distribution. The GABP••••–overexpressing clones 3 and 10 were grown for 36 hours in the presence and absence of the inducer *Doxycycline*. The cells were fixed, PI-stained and counted by FACS. For negative control the parental cells S2-34 were used. The percentage of cells in each cell cycle phase (+/- *Doxycycline*) is denoted in tables below the graphs. The sum of the phases' percentile values does not equal hundred due to loss of events outside the marked areas. No particular pattern in the cell cycle phases distribution is observed. Usually the differences between the single points fall well into the error bars, indicating lack of effect from the GABP•••• overexpression.

This figure depicts average data from twelve PI-staining experiments. The percentile distribution for each pair of samples is determined several times (between 3 and 5 times) independently and averaged. Furthermore, the data sets for all similar samples are averaged additionally.

Cell cycle distribution for sub-confluent cultures cultivated in the presence or in the absence of *Doxycycline* was determined by Propidium Iodide (PI) staining followed by FACS analyses. Two NIH-3T3-GABP•••• clones were used for the experiment (••••-3 and 10) and NIH-3T3-S2-34 served as negative control. No major differences in the cell cycle phases' distribution with or without induction were found (Fig. 2.20).

Results

To consolidate these results, we analyzed mRNA expression for all major cyclins under the same differential conditions. In the experiment three clones of the cells NIH-3T3-GABP•·· (35, 3 and 8), expressing exogenous GABP•·· to different extend were grown in presence of inducer (*Doxycycline*) for 24 and 72 hours and from these cultures, together with a negative control (lacking induction), cell extracts were prepared and used for the assays. No visible differences among the three time points for each clone were detected, indicating that the observed change in the cell proliferation speed during the time of elevated GABP•·· expression is not due to changes in the transcriptional regulation of cyclins (Fig. 2.21).

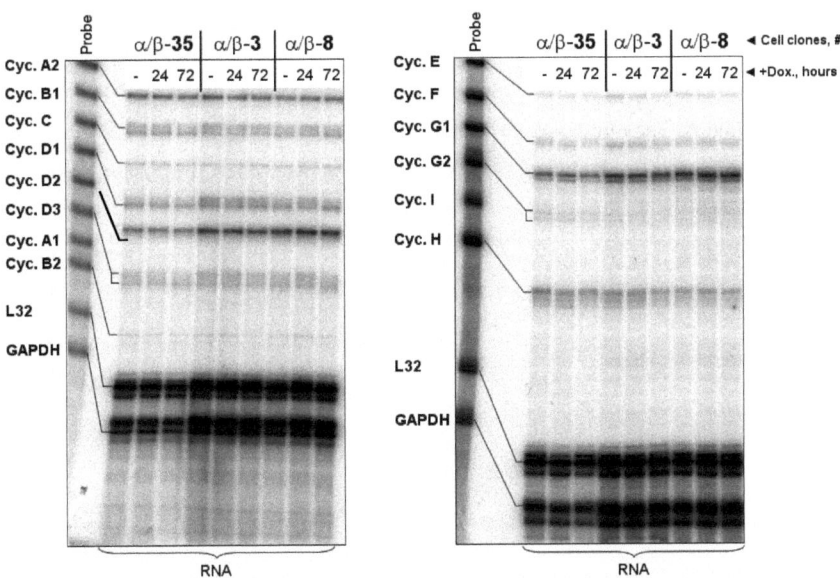

Fig. 2.21: Increased GABP•·· expression does not influence the transcriptional regulation of the cell's cyclins: The GABP•··–expressing clones •··–3, •··–8 and •··–35 were grown for 24 and 72 hours in presence of the inducer *Doxycycline*, RNA was extracted and used for RNase protection assay. For negative controls the clone •··–35 was used, expressing endogenous levels of GABP•·· and the clones •··–3 and •··–8 not treated with inducer. RNase-protection kits for mouse cyclins used: mCyc1 and mCyc2.

This leaves the last possibility – that the observed culture growth speed reduction reflects some apoptotic processes resulting from expression of more than endogenous levels GABP•··.

Results

2.9. Excessive Expression of GABP•·· Increases Apoptotic Processes in NIH-3T3 Cell Cultures

2.9.1. Increased Sub-G$_1$ Cell Cycle Population Indicates Increased Apoptosis in GABP•·· Overexpressing Cell Cultures

The presence of so-called Sub-G$_1$ population is an indicator of apoptotic processes in cell culture. During conventional cell-cycle analyses of adherent cells this fraction is usually lost during the washing/trypsinization/cell harvesting steps.

To detect sub-G$_1$ population, growth medium from a set of NIH-3T3 cell clones, cultivated in the presence or absence of *Doxycycline* was collected, cellular debris were isolated by centrifugation and after PI staining were analyzed on FACS (Fig. 2.22).

Cell cultures with inducible expression of GABP•·· clearly showed increased numbers of sub-G$_1$ events after addition of *Doxycycline* while opposite effect was observed in control cells. In addition, absolute values of sub-G$_1$ increase correlated well with the level of GABP•·· induction in analyzed clones, suggesting apoptotic processes resulting from increased GABP•·· expression.

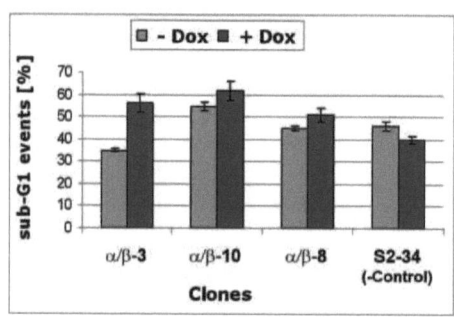

Fig. 2.22: Sub-G$_1$ peak events in GABP•·· overexpressing cells show increased values. The GABP•··–overexpressing clones 3, 8 and 10 were grown for 36 hours in presence and absence of the inducer *Doxycycline*. The dishes were shaken by tapping, the culture media was collected and the cells and cell debris it contains were harvested by centrifugation, fixed, PI-stained and counted on FACS. For negative control the parental cells S2-34 were used. The percentile value of the Sub-G$_1$ events out of the total cell population is plotted on the graph. Note that only the bodies, freely floating in the cell media were collected. Increase of the sub-G$_1$ event values in presence of *Doxycycline* for all GABP•·· overexpressing clones was evident, while the value of the negative control rather decreases. The figure interprets data from one of several similar experiments. Error bars represent Standard Deviation.

2.9.2. Examinations of the Caspase Pathways Reveal Elevated Expression and Activation of Some of Their Key Members

Initiation and execution of apoptosis relies on a complex network of caspases – cysteine proteases which also are critical for a normal development of several cell lineages (Fig. 2.24).

Results

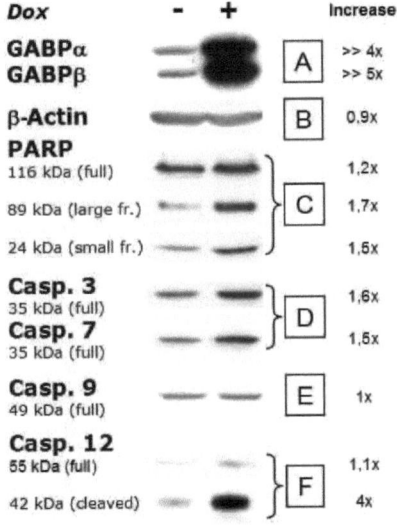 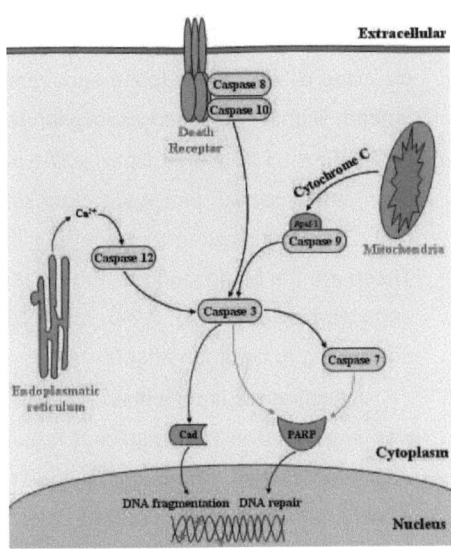

Fig. 2.23: Activation of caspases after GABP·‑‑ overexpression. NIH-3T3-GABP·‑‑-3 cells were grown for three days with inducer *Doxycycline*. Whole cell extracts were utilized for Western blots. For negative control non-induced cells were used. **A,** Expression of both GABP subunits is strongly increased upon induction (off-scale). **B,** ·-Actin control shows equal loading (valid for all shown antibodies). **C,** PARP expression and cleavage is upregulated by the higher GABP expression. **D,** Caspase-3 and caspase-7 expression is also upregulated. **E,** Caspase-9 expression is not changed. **F,** Caspase-12 expression is elevated and its cleavage is strongly upregulated.

This figure is compiled from different exposures and antibody treatments of the same membrane, from a single experiment. The data are confirmed with additional experiments. The strength of the signal is determined by densitometry.

Fig. 2.24: Major scheme of the three main caspase pathways:
- Death receptor pathway, involving activation of caspases -8 and -10
- Mitochondrial pathway with key member caspase-9
- Endoplasmatic Reticulum pathway, which is controlled by caspase-12

Each of them is activating caspase-3, which in turn activates caspase-7 and both cleave PARP.

Initiation of apoptotic processes results in activation of one or several caspase cascades. This is reflected in an appearance of cleaved, activated forms of caspases and increased cleavage of effector caspase substrates, such as poly ADP-ribose polymerase (PARP). Indeed, slightly increased expression level and increased cleavage of PARP was evident in GABP•••3 cells after *Doxycycline* treatment (Fig. 2.23.C).

As expected, increased expression of effector caspases, -3 and -7, was detected (Fig. 2.23.D) indicating that some of three major caspase pathways must be activated. Those are the Mitochondrial-, the Death receptor- and the ER (Endoplasmatic Reticulum) -induced pathways (Fig. 2.24), which could be activated by different classes of stimuli but are leading to common effects.

Caspase-9 is an initiator caspase of mitochondrial pathway. As GABP•••complex was implicated in transcriptional regulation of several genes involved in mitochondrial respiratory chains, one might assume that deregulation of these genes might trigger the mitochondrial apoptotic pathway. However, Western blot analyses did not reveal any effect of increased GABP••• expression on caspase-9 (Fig. 2.23.E) arguing against the involvement of mitochondrial pathway in apoptosis.

Caspase-12 is responsible for the initiation of apoptosis via the ER pathway, which is thought to be independent from mitochondrial and death receptor pathways. In response to ER-stress, caspase-12 is leaving the ER and transferring the signal to effector caspases.

Initially unexpected, Western blot analyses revealed slightly increased expression of pro-caspase-12 and very strong induction of activated caspase-12 in cells with higher GABP••• expression (Fig. 2.23.F). This suggests that either increased GABP••• expression leads to ER-stress *per se* or that GABP••• is upregulating other signaling molecules which in turn directly or indirectly lead to activation of caspase-12.

2.9.3. Increased Expression of GABP•••*Per Se* Does Not Induce ER-Stress

We have considered the possibility that the expression of extra amounts exogenous proteins (namely GABP• and GABP• subunits) in the cell may result in ER Overload Response (EOR), depletion of cellular resources or Unfolded/misfolded Protein Response (UPR). That's why experiments, designed to answer that questions were performed.
Examination of Coomassie/silver stained protein gels (data not shown) and Ponceau stained membranes (Fig. 2.27) did not reveal any *Doxycycline*-inducible protein band in whole cell or nuclear extracts from •••-3 cells. As shown before, overexpressed GABP•

and GABP• are exclusively located within the nucleus, and an increased staining of any cytosolic compartment and especially of the granulated ER was not detected (Fig. 2.11, Fig. 2.25 and Appendix). In addition, a weak suppressive effect of GABP• •• was evident in NIH-3T3 clone with the lowest level of GABP• •• induction (i.e. • ••-42, see Fig. 2.14). Altogether these findings are ruling out the possibilities that too high expression level of GABP• •• is inducing EOR and/or results in a depletion of cellular resources therefore leading to ER-stress with consequent activation of caspase-12.

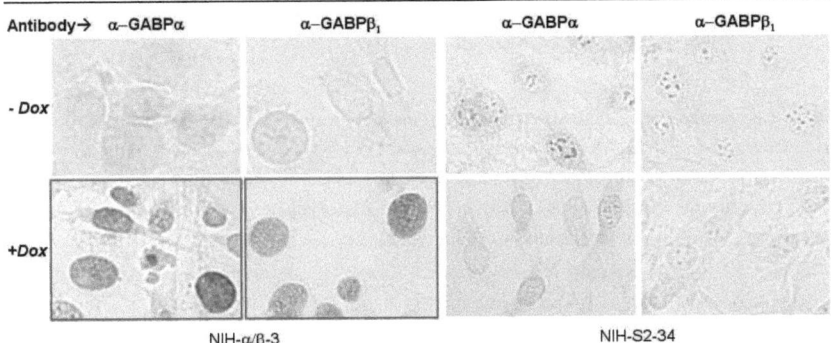

Fig. 2.25: Lack of cytoplasmatic accumulation of the conditionally overexpressed GABP• and GABP•. The intracellular localization of GABP• and GABP• is visualized by cytostaining of the cell clone NIH-3T3-GABP• ••-3 compared to the parental NIH-3T3-S2-34 cells. Both cell types are stained with anti-GABP• and anti-GABP• ₁ antibodies and half of them are induced with *Doxycycline* for 36h. No DNA staining is performed on the shown samples to avoid masking of the protein signal. Predominantly nuclear localization of the both subunits is observed. In case of overexpression the nuclear signal is many times stronger but still – increase in the cytoplasmatic signal can not be detected.
In total five cytostains were performed, at different times and in various combinations of staining agents/antibody. Pictures in magnifications 10x and 40x were taken, from which only the latter were used in the figures. The results are consistent throughout the different experiments.

The GABP• •• expression vector used for stable transfections in our study does not encode any intentionally added 'tag'- or fusion- sequences which eventually might result in an inappropriate folding of expressed proteins and as a consequence – induce UPR. However, as noted above (Chapter 2.5), in this vector which was used to create NIH-3T3 clones a single point mutation in the GABP• -coding region was detected. To exclude the possibility that the abovementioned 6 amino acid extension of C-terminus might result in a misfolding of GABP• and induce UPR, NIH-3T3 cells were transiently transfected with the verified wild-type versions of GABP• and GABP• expression vectors. Western blot analyses revealed a strong activation of caspase-12 in cells when both GABP• and GABP• were co-transfected (Fig. 2.26.). At the same time no increase in caspase-12 activity was detected when GABP• was overexpressed alone. Only slight increase was observed in GABP• transfected NIH-3T3 cells. This represents very well matched correlation with the previous data, observed on the stably transfected NIH-3T3-GABP•••-3 cells.

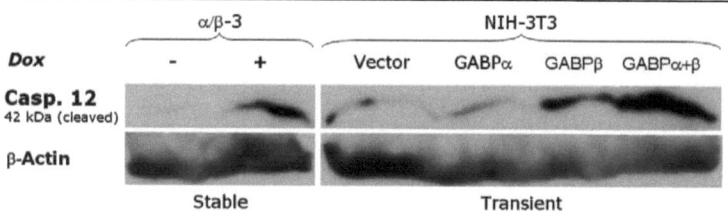

Fig. 2.26: Upregulation of caspase-12 caused by transiently transfected wild type GABP. Two independent transient transfections on NIH-3T3 cells (with efficiencies 40-50%) were performed, in which the samples were co-transfected with the plasmid pMAX-EGFP as a transfection efficiency marker, together with the vectors: LX (empty), LX-GABP•, LX-GABP• and LX-GABP• plus LX-GABP•. The signal is visualized by Western blot. Loading control used is •-Actin, control samples are NIH-3T3-GABP•••-3 cells without induction and with induction with *Doxycycline* for three days. In both cell types the highest amounts of active caspase-12 are present when GABP• and GABP• are overexpressed simultaneously, confirming the data from Fig. 2.23.

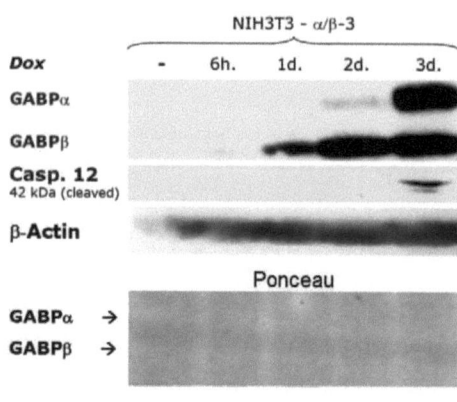

Fig. 2.27: Upregulation of caspase-12 caused by conditional overexpression of GABP• and GABP•.

Upper panel: Western blots from NIH-3T3-GABP•••-3 cells, induced with *Doxycycline* for up to three days. Note that the caspase-12 signal is higher at the time point when both GABP• and GABP• are simultaneously overexpressed in their most prominent amounts. Loading control – •-actin. At least three independent, Western blot-visualized, GABP induction time course experiments are performed. Here, part of the best image is depicted.

Lower panel: Ponceau staining of the same membrane after the blotting. No increase in any of the bands in the area of GABP• or GABP• is visible by naked eye.

These results indicate that the stably transfected overexpressed GABP•••subunits *per se* are not resulting in ER-stress. Instead, caspase-12 in all likelihood is activated through the modulation of direct or indirect GABP••• target genes.

3. DISCUSSION

One of the preferred ways to study an unknown protein is to modulate its expression level. The most popular methods are knock-down, knock-out and overexpression. This usually results in alterations of the protein's function. Similar effect can be achieved also by introducing abnormalities in the protein's primary structure via creation of mutations in its coding sequence. Such desired mutations can lead to an expression of inactive or constitutively active versions of the protein. These changes in the expression level and/or activity usually result in distinct phenotypical traits. The changes do not have necessarily to be direct though – e.g. an overexpression could not only increase the intensity of an existing phenotypic trait, but also decrease it, similarly to the expected decrease achieved by the knock-down method. This can be caused by a potential negative regulatory function of the overexpressed protein, executed over the regulation of a second, trait-forming protein.

3.1. Efficient Downregulation of GABP• Expression by RNAi Could Be Achieved Only Transiently

Several groups with different success have tried to establish *in vivo* and/or *in vitro* systems with deficient GABP• expression. We detected very early lethality of GABP•˙⁻ embryos (before E3) and speculated that GABP••• complex, which is highly expressed in embryonic stem (ES) cells, might be required for self-renewal of these cells (A. Avots, 2002, unpublished). This embryonic lethality and the proposed involvement of GABP were confirmed later (Ristevski et al., 2004 and Kinoshita et al., 2007). Problems in early developmental regulation were also believed to take effect, attributed to presumable involvement of GABP complex in cell proliferation and/or differentiation processes, which are essential for the early embryogenesis. W. J. Leonard's group took advantage of available library of gene-trap ES cell clones and generated mice in which reporter gene was inserted into the intron 5 of the Gabp• allele. In the embryos that are homozygous for Gabp•-"trapped" alleles (Gabp•$^{tp/tp}$), GABP• expression was greatly diminished but still detectable. This minor expression enabled the Gabp•$^{tp/tp}$ embryos to survive until E12,5 to E14,5 thus allowing analysis on the role(s) of GABP *in vivo* and *ex vivo* (Xue et al., 2004). Finally, real conditional GABP• KO mice were established in Rosmarin's lab and successfully used to identify GABP•••-dependent but E2F-independent pathway of

Discussion

cell proliferation. Once created though, GABP• KO cells were unable to proliferate further because the entry into S phase was prevented and more investigations on the connection between GABP and the cell culture's growth speed could not be performed (Yang et al., 2007).

RNA interference decreases GABP• mRNA levels and hence – the GABP• protein in cultured cells. Therefore we are able to circumvent the difficulties, arising from the early lethality of GABP•$^{-/-}$ embryos in GABP• KO mice. This method is faster and in comparison with the conditional knock-out can be more easily used, particularly in cell culture. Indeed, we were able to achieve impressive decrease in GABP• expression in U2OS cells after transient transfection with several plasmid constructs bearing siRNA-expression cassettes directed against GABP•. Residual expression level is crucial though, as small decrease might influence some but not all functions of a given transcription factor, while too strong decrease can completely eliminate the function expected to study. For instance, GABP• levels were only marginally influenced in all examined tissues from heterozygous GABP•$^{+/-}$ knock-out mice (Ristevski et al., 2004). As a consequence, such animals or cells show no difference in their physiological properties in comparison to the wild type animals/cells. Another extreme condition is when the GABP• is completely knocked out. In this case the subjected cells completely stop their proliferation (Yang et al., 2007), so any further investigation is becoming problematic. In our experimental settings we aimed to achieve considerably large but not complete decrease of GABP• expression.

The U2OS cells are human osteosarcoma derived cell line which was intentionally chosen for experiments with siRNA constructs to establish procedure to target both human and mouse GABP• gene. GABP• genes in human and mouse are highly homologous, therefore we screened and successfully identified regions identical both in the human and mouse genes which at the same time fulfilled criteria for efficient siRNA targets. Transient transfection of U2OS cells with several constructs, bearing various siRNA-forming sequences indeed revealed several promising siRNA targets and with two of them we observed almost complete absence of the GABP• protein after three days of selection (see Fig. 2.1). However after transfer of the most efficient siRNA expression cassette into the retroviral vector and subsequent transient transfection of the same cells with this new construct we detected only ten fold reduction of GABP• expression. This finding was attributed to the larger molecular mass of the retroviral vector (more than twofold) or more likely, to the organization of retroviral transcriptional units. Nevertheless,

Discussion

the achieved downregulation seemed to be sufficient for the intended stable long-term downregulation of GABP• expression.

However, observed knock-down effect was relatively short-lasting and detectable during the first week after transfection ("transient transfection" period). Later, after three or more weeks of selection ("stable transfection" stage) the knock-down effect was decreased to values little differing from the mock-transfected control (from 1,15 to 1,3 fold). Such rates, although detectable and suggesting that knock-down processes might be still going on in these cells, were not as strong as to be desired. We generally anticipated that only a decrease, higher than 15% or even 30% of the protein amount would result in detectable and more importantly – reliable and clearly visible change, especially when studying a regulatory protein with as many diverse functions as the GABP complex. Thus, the observed results were leading to only possible conclusion that the GABP knock-down, higher than 15-30% was not tolerated by the selection process in long-term manner. This indirectly, but strongly supports the assumption that GABP has a crucial function for maintaining the normal growth of the cell culture.

Expression levels of GABP• seem to be tightly regulated in mice. Comparison of many different cell types derived from heterozygous (Gabp•$^{+/-}$) and homozygous (Gabp•$^{-/-}$) animals shows no detectable differences in the amount of that protein for each particular cell type (Ristevski et al., 2004). This indicated that it might not be possible to achieve lasting substantial down-regulation of GABP• expression in U2OS cells.

The diminishing of the GABP• knock-down efficiency was attributed to a possible negative selection against cells with low level of GABP• expression. In such a case cells expressing GABP• below certain threshold levels would be unable to proliferate further or their speed of proliferation would be decreased, leading to continuous decrease in the total number of such cells in the culture. This hypothesis was supported by previously existing data indicating involvement of GABP in the cell cycle regulation. As revealed by various authors GABP is essential for the transcription of Skp2 (Imaki et al., 2003), retinoblastoma gene (Savoysky et al., 1994; Shiio et al., 1996 and Sowa et al., 1997), E2F1 (Hauck et al., 2002), DNA polymerase • (Izumi et al., 2000) and thymidylate synthase (Rudge and Johnson, 2002) – all of them important regulators of the cell cycle progression from G_1 to S phase. More confirmations come from publications showing that GABP is required for G_1/S cell-cycle progression (Yang et al., 2007) and that the knock-down of GABP arrests cell cycle progression through G_1/S phases by indirect regulation of the cyclin-dependent kinase inhibitor p27 (Crook et al., 2007). In addition, GABP was implicated as a positive regulator of the aurora A gene transcription and therefore it is involved in the regulation of the G_2/M checkpoint of the cell cycle (Tanaka et al., 2002).

A possible molecular mechanism to be considered is selective silencing of H1 promoter within the expression cassette. This suggestion comes out from an interesting observation, when in the transfected cells a GFP fluorescence marker was also present. In such a case the cells were either co-transfected with the siRNA-bearing plasmid pRS10 together with the GFP-expressing plasmid pMSCV-Puro-GFP, or transfected with the combined plasmid pRS10G. The stably transfected cell population after three weeks of Puromycin selection retained almost the same percentage of green fluorescing cells as in the transiently transfected culture after only three days of selection (more than 90% in both cultures). Obviously all of the surviving cells after selection must carry the resistance against Puromycin, therefore the negative selection could be explained only by selection against the GABP• siRNA-expressing cassette but not against the cassettes, bearing the selection and/or visualisation markers, regardless of whether they are positioned in the same plasmid or not. This strongly argues against complete silencing of the retroviral transcriptional unit. However we can not formally exclude selective silencing of H1 promoter which was used to express siRNA transcripts.

Insuring that the described lack of long-lasting knock-down effect is not due to an experimental glitch, a variety of alternative attempts to achieve it were conducted. They include testing of several siRNA expression vectors, targeting different sequences within the GABP• gene, changing orientation of promoter/siRNA cassette and testing RNA interference in other cell types. All these experiments resulted in a similar, very low level of GABP• downregulation in stable cultures.

3.2. Constitutive GABP• Overexpression Speeds Up Cell Proliferation

Since the downregulation of GABP• resulted in a likely negative selection against the cells with low GABP• expression level or in a selective silencing of H1 promoter we assumed that an elevated expression will not face similar problems. A selection against the GABP• •• highly expressing cells was not expected, as positive influence on cell proliferation speed was anticipated. Recently it was shown that GABP• overexpression positively influences the self-renewal of ES cells even in the absence of LIF (Kinoshita et al., 2007). Proposed mechanism suggests increased expression of Oct-3/4 via downregulation of its repressors Cdx-2, Coup-tf1 and GCNF. It is also supposed that the

Discussion

increase of the GABP• gene dosage in the Ts65Dn segmental trisomy mouse model of Down Syndrome may play a role in DS pathologies in tissues where GABP• protein levels are elevated (brain and skeletal muscle). However, according to other sources increase in the protein level of GABP• could not be detected when its mRNA was overexpressed in human DS fibroblast cell lines or after transfection of NIH-3T3 cells (O'Leary et al., 2004). This suggested tight regulation of the GABP• protein levels, confirmed also by the group of P. J. Hertzog. They created $Gabp\bullet^{+/-}$ mice which demonstrated unaltered protein levels in the panel of tissues examined (Ristevski et al., 2004).

Appropriate cell model had to be used to investigate the role of GABP on the growth speed of mammalian cell cultures. Considering some specific preferences, the mouse embryonic fibroblast cell line NIH-3T3 was chosen to proceed with as well suited for our general purposes. These cells represent a continuous cell line, without a limit to the number of generations that they can propagate. They are still relatively close to the primary embryonic fibroblasts from which they were derived and are morphologically similar to their primary counterparts. As anchorage-dependent adherent cells they propagate as a monolayer attached to the culture vessel thus enabling easier transfection, selection and especially cloning. NIH-3T3 cells are intensively used for cell cycle regulation and transformation studies and, in addition, they were used as a host for Tet-system before.

First step to achieve elevated expression of GABP••• was to express extra amounts of the GABP• subunit which is responsible for the transport, nuclear localization and transcriptional activity of the whole GABP••• complex. Thus, the aim was to achieve complete nuclear localization and mobilization of the total available amount of cellular GABP• from the cytosol. This idea was supported by data indicating that overexpression of GABP• increases *Skp2* promoter activity (Imaki et al., 2003) and with our data indicating the presence of cytosolic GABP• fraction in several cell types (see Fig. 2.3). Stably transfected, constitutively overexpressing, polyclonal NIH-3T3 cells expressing high amounts of GABP• were created. Interestingly, overexpression was greatly diminished when the cells were let to reach confluence. This effect can be due to the supposed decrease of the total protein production induced by the contact inhibition, because of the specific down-regulation of GABP• protein level in non-proliferating cells or, more likely these conditions result in a decreased activity of the promoter used in the expression vector. The increase of the GABP• expression did not cause an increase of

Discussion

the total amount of GABP• protein though, nor was detected additional recruitment of GABP• from the cytosol into the nucleus, suggesting that there is no substantial excessive cytosolic fraction of GABP• in these cells. The quantities of the cytosolic and nuclear GABP• fractions didn't show any detectable change probably due to the tight regulation of the endogenous GABP• subunit (see chapter 1).

GABP•-overexpressing NIH-3T3 cells were growing substantially faster (20%) than cells transfected with the empty vector only. This preliminary experiment indicated that GABP••• complex has an influence on the proliferation speed of mouse cells and prompted for more detailed investigations. To achieve this goal we established a system for conditional simultaneous expression of exogenous GABP• and GABP• subunits in NIH-3T3 cells.

3.3. Conditional Expression of Excessive GABP• /• in NIH-3T3 Cells Reduces Cell Proliferation Speed

In order to achieve a proper operation of the exogenously expressed GABP subunits and to provide better negative controls, a more sophisticated experimental approach was adopted. We aimed to conditionally and simultaneously elevate the expression of the both GABP• and GABP• subunits.

The multi-step DNA-cloning (see the Appendix), two consecutive transfection and single-cell cloning steps yielded a panel of cell clones with inducible expression of exogenous GABP. Each clone was able to express GABP• and/or GABP• subunits at a slightly different level, in a range from just a little elevated expression till as much as ten- to fifty fold increase for GABP• and GABP• respectively. Individual selected clones were tested for inducible expression of both GABP• and GABP• (Fig. 2.10) and their DNA-binding properties *in vitro*. The DNA binding activity of GABP••• *in vitro* correlated well with their expression levels (Fig. 2.12). Overexpression, nuclear targeting of GABP• and the ability to establish ••• and ••• bonds (forming respectively dimers and tetramers) was detected by cytostaining and EMSA.

Some sources (Yang et al., 2004) state that both GABP subunits are localized both in the nucleus and cytoplasm. Our observations though, show that in NIH-3T3 cells both of the subunits exert predominant to near exclusive nuclear localization (see Fig. 2.11). In other occasions, in spite of detected overexpression of GABP• mRNA in human Down

Discussion

syndrome (DS) fibroblast cell lines, neither in these cells nor in GABP• -transfected NIH-3T3 cells differences in protein levels could be found (O'Leary et al., 2004). However, our own data show that in GABP• -transfected NIH-3T3 cells elevated amounts of the protein are easily detectable.

The activity of the GABP complex itself was proven by the lack of dominant negative effect in a transient transfection assays using luciferase reporter constructs under the control of multiple GABP binding sites (data not shown).

Trough GABP• •• overexpression, the proliferation speed of each individual clone was decreased, where the degree of decrease correlated with the strength of overexpression. At first sight, these results seem to contradict with the results from the previous chapter, where overexpression of GABP• alone increased the proliferation speed of the cell culture. However, when evaluating the divergence in experimental results, two important facts must be taken into account. The first one is manifested by transfecting at one case only the functionally incomplete GABP• and at another – the full functional set GABP• plus GABP•. By using GABP• alone, mobilization of the total GABP• amount from the cytosol and its complete re-localization into the nucleus was aimed. Although such effect was not clearly observed by the used method of visualization, we considered the possibility that the exogenous GABP• was able to nuclearly transfer sufficient amounts GABP• for at least partial increase of its concentration, hence positively altering the culture growth speed. Taking into account the abundance of various functions attributed to GABP, we can not rule out the possibility that in different concentrations it regulates completely different cell processes, leading to dissimilar outcomes concerning the cell growth speed. Such unparalleled outcomes can be only supported by the even more important second difference in the experimental settings. Namely, this is the fact that GABP• was transfected in a manner with stable, constitutive expression in contrary to the inducible expression of the GABP• •• construct. A constitutive expression gives the opportunity of an exogenous transcription factor to function during the transfection and selection stages, while the inducible expression is switched on only at desired time points. Functioning during the selection stage of an experiment is highly undesired, when investigating properties, connected with the cell growth speed. In this case, the selection favors only the faster dividing cells, eliminating all slow growers. When the transcription factor confers negative or positive influence on the growth speed the first would be diminished or completely obliterated while the second would be supported and enhanced. These logical assumptions are matching completely our case, where constitutively expressing GABP• cells are demonstrating increase in their growth speed, opposite to what is observed by the cells, inducibly expressing GABP•••.

It has to be noted that a single point mutation at the very end of the GABP• coding sequence of the transgene was detected *post factum*. This mutation transformed the stop codon (TGA) into Gly codon (GGA) and resulted in a six amino acid long C-terminal extension of GABP• subunit. To the best of our knowledge, this C-terminal extension did not affect any described property of GABP•.

In addition, transient overexpression of wild type GABP•••, in NIH-3T3 cells showed identical effects to those, demonstrated by the conditional overexpression of GABP••• bearing the 6 additional amino acid residues (See next sub-chapter).

3.4. Elevated Expression of GABP• •• Induces Caspase-12 – Elicited Apoptosis

Flow cytometry analysis showed no difference in the size of cells from cell cultures with elevated GABP amounts. Therefore we were able to measure the cell proliferation rate by comparing the cell mass only. These experiments revealed the interesting observation that cell culture growth speed decreases as a consequence of the increased GABP expression and importantly, that elevated expression levels correlated well with the cell proliferation speed reduction. It was also observed that a prolonged exposure of the cells to higher GABP levels further decreases their proliferation rate. This process was completely reversible after withdrawal of *doxycycline* and following return to the normal GABP expression levels. Growth speed of such cultures returned to normal within only three days. In addition, we observed a specific growth pattern of the slower proliferating cells (colony-like formations). These observations led to the necessity to examine likely mechanisms responsible for those effects, including investigation of the cell cycle distribution and apoptosis.

Staining of the cellular DNA set with Propidium Iodide and flow cytometry was used to determine the cell cycle distribution. No differences were detected between the cultures of identical clones grown under normal and elevated GABP• •• expression. In both groups the pattern of cell cycle phase distribution was practically identical and had wild type characteristics. The cell numbers in G_1, S and G_2 phase were indistinguishable for both expression groups. The majority of the events were placed in G_1 phase, followed

Discussion

by those in G_2 and a minor amount in S phase. These results indicate that endogenous levels of GABP•·· are sufficient for a proper cell-cycle progression of NIH-3T3 cells and suggest that further increase of these levels does not result in detectable changes of this progression.

To confirm this conclusion, transcription of cyclin genes was checked at several time-points after Doxycyline induction of transgenic NIH-3T3 cells with different induced levels of GABP•·· proteins (clones 35, 3 and 8). RNase protection assays did not reveal any changes in the transcription level of the mouse cyclins A1, A2, B1, B2, C, D1, D2, D3, E, F, G1, G2, H and I after elevation of GABP•·· amounts (Fig. 2.21), indicating that other process(es) are responsible for the observed change in the growth speed of the cell cultures expressing additional GABP.

Intriguingly, examination of DNA amount profiles during the cell cycle clearly indicated the increase of sub-G_1 population after elevation of GABP•·· expression levels suggesting decline of the cell culture's health, presumably caused by apoptosis (Fig. 2.22). The extent of sub-G_1 population increase correlated well with the absolute levels of GABP•·· expression in individual NIH-3T3 cell clones. Importantly, this effect could not be attributed to a toxic effect of the inducer, as Sub-G_1 population was rather decreased when *doxycycline* was given to control cells which do not express exogenous GABP•··. When analyzing the results from Fig. 2.22, must be noted that the plotted data are derived from objects, floating in the culture media only – the monolayer of cells was not disturbed during the collection process. This implies that the ratio Total events/Sub-G1 events depends not only on the cell debris, but also on the loose cells (dividing cells, apoptotic cells). Loose cells are presumably present in higher numbers at apoptotic cultures, decreasing their Sub-G1 percentage. Therefore, this data must not be directly bound and compared with data, obtained from other experiments.

Undergoing apoptotic processes were undoubtedly confirmed by the detection of numerous activated caspases and their products. It is noteworthy that for some of the caspases also non-apoptotic functions were described, especially in cellular proliferation and differentiation (Schwerk and Schulze-Osthoff, 2003). These functions are in several cases mimicking and very much related to their apoptotic functions, where the major difference is the level of caspase activation. It is believed that the caspase activity level represents the difference between the apoptotic and non-apoptotic phenotype. In our experiments, in similar manner a mild increase of the activity of several caspases was observed, causing also not full-scale apoptosis, but decrease of the growth speed of the

Discussion

affected cultures. Initially, increase in the cleavage of poly ADP-ribose polymerase (PARP) was observed on Western blot. PARP is a substrate of the downstream effector caspases caspase-3 and/or caspase-7 and appearance of a cleaved form of PARP is an indication for increased caspases activity. Following examinations detected an increased expression and activation of those two caspases suggesting that the whole caspase-regulated apoptotic network is activated. To identify which pathway is triggering the activation of the effector caspases in particular, specific members of two out of the three main paths were analyzed. Caspase-9, crucial for activation of the mitochondrial pathway, was proved to maintain the same expression levels throughout the process of increasing the GABP expression. In contrast, the expression level of caspase-12, responsible for the activation of the Endoplasmic Reticulum (ER) pathway, was substantially upregulated. In addition, increased cleavage/activation of caspase-12 suggested that ER pathway is a major if not only reason for the initialization of apoptotic processes after elevation of GABP• •• expression. Therefore the other major apoptotic pathway – Death Receptor pathway (specified by caspases -8 and -10) as well as the diverse interconnections between the pathways were not investigated.

In our experimental set, we elucidated clues supporting the notion that overexpression of GABP *per se* does not induce ER-Overload Response (EOR). In most of the clones used, the GABP• subunit was expressed about 10-20 fold and GABP• – about 50 fold over the endogenous levels. As the transcription factors which are regulatory proteins are expressed at relatively low levels in comparison to other cellular proteins (e.g. structural), the probability for EOR is miniscule. In addition, in several other cell types GABP is normally expressed in very high amounts, indicating existence of tolerance for increased GABP levels. Particularly GABP is more abundant in liver, muscle, ESCs and hematopoietic cells, although not in fibroblasts as NIH-3T3 cells (LaMarco et al., 1991 and Brown et al., 1992). Therefore, we suggest that the observed overexpression is not able to overload cellular protein synthesis machinery. This proposal is supported by the lack of observable evidence for GABP accumulation in ER. The cytostains, performed for detection of the intra-cellular GABP distribution proves its nuclear localization and lack of accumulation in any ER-related cub-cellular structures (ER or Golgi apparatus) (see the Appendix).

In addition, to prove that activation of caspase-12 resulted from the elevated amount of GABP and to exclude any association with C-terminal extension of exogenous GABP• (see chapter 3.3.), additional transient transfection experiments were conducted

Discussion

using several independent and verified expression vectors. It should be noted that GABP•‧‧ expression vectors used in this study do not encode any N- or C-terminal 'tag' sequences which theoretically might change the conformation of the exogenously expressed proteins and result in unfolded protein response, specified by ER stress and activation of caspase-12. As expected, large amounts of overexpressed GABP subunits were detected in transiently transfected cells. Distinct increase in the activation of caspase-12 was observed when both -• and -• subunits were overexpressed simultaneously. This proves the requirement for •‧‧ coupling to ensure proper activity of the complex. When only the expression of GABP• subunit was elevated, caspase-12 was cleaved to lesser extend while higher expression of GABP• alone showed no influence on caspase-12 activation. These data support the assumption that the amount of endogenous GABP• subunit exceeds that of the GABP•, so the addition of -• subunit increases the total amount of transcriptionally active complexes. Thus, the specificity of the GABP effect on caspase-12 was proven along with the reliability of the conditional GABP•‧‧ overexpressing NIH-3T3 cells and the functional integrity of the transgenic proteins.

The question whether the activation of caspase-12 by elevated levels of exogenous GABP•‧‧ reflects a normal physiological mechanism of caspase-12 regulation remains unanswered. The available data concerning the regulation of caspase-12 expression and activation are rather scarce and clear connection with a potential GABP involvement with these processes can not be easily established. It is known that in mouse constitutive expression of the caspase-12 protein is restricted to certain cell types, such as epithelial cells, primary fibroblasts, L929 fibrosarcoma cells (Kalai et al., 2003), renal proximal tubular epithelial cells, high levels expressed in muscle, liver and kidney and moderate levels in brain - in cortical neurons, Purkinje cells, brainstem neurons and olfactor neurons (Nakagawa et al., 2000). In fibroblasts and B16/B16 melanoma cells, caspase-12 expression is stimulated by IFN•• but not by IFN•• or ••. The effect is increased further when IFN•• is combined with TNF, lipopolysaccharide (LPS), or dsRNA (Kalai et al., 2003). Studies in caspase-12–deficient mice suggested that the protein specifically plays a major role in ER stress–induced apoptosis and in the development of Alzheimer's disease (Nakagawa et al., 2000). Since then, several other reports have linked processing of caspase-12 to ER stress–induced apoptosis (Rao et al., 2001, 2002; Diaz-Horta et al., 2002; Morishima et al., 2002; Fujita et al., 2002). In addition, caspase-12 seems to be involved in apoptosis induced by viral infections (Bitko and Barik, 2001;

Jordan et al., 2002) or by serum starvation (Kilic et al., 2002). Recently, several reports linking the ER-apoptotic pathway and caspase-12 activation with the activation of other apoptotic pathways became available. The mitochondrial pathway plays significant role in ER stress-induced apoptosis in MEFs (Shiraishi et al., 2006) and it appears that these two pathways reinforce each other during the apoptotic process (Sanges and Marigo, 2006). However, our results suggest that this pathway does not contribute to the final activation of caspase-3, observed in cells with extra GABP• •• amounts.

Special attention has to be paid on the emerging link between the caspase-8 and caspase-12 pathways. It was observed that TNF• -mediated apoptosis in HL-1 cardiomyocytes follows the caspase-12 apoptotic pathway that involves calpain (Bajaj and Sharma, 2006). Calpain activation was found to mediate caspase-12 activation also in 7-Ketocholesterol-induced apoptosis where, interestingly caspase-8 activation was also observed (Neekhra et al., 2007). GABP is known to enhance the transcription of TNF-• (Tomaras et al., 1999• and together with AP-1 is required for initiating Fas gene transcription (Li et al., 1999), both of which are known to be involved in the induction of apoptosis via the Death receptor pathway. This implies that a possible link between these two pathways has to be also considered.

The exact molecular mechanism of the GABP induced caspase-12 cleavage upregulation though, remains unclear and is an attractive subject for further studies. It is not investigated whether GABP directly regulates the cleavage of pro-caspase-12 to caspase-12, is it stimulates some of the upstream factors involved in the ER stress response, triggers the TNF receptor/Fas apoptotic pathway or acts through another, yet unidentified mechanism.

Finally, an important question about the physiological relevance of the caspase-12's pathway induction must be attended. As we have seen, a minor gain in the caspase-12's expression is coupled with major increase in its activation, leading to rather weak activation of the downstream caspase-3. The end result from this process is observed as a phenotypical trait in the form of mild increase in the percentage of culture's apoptotic cells. This is the effect of GABP overexpression, present in the model cell culture of NIH-3T3 cells, which normally do not express large amount GABP and are belonging to already differentiated fibroblast cell type. Assessing the question on a more physiological background, we must compare this effect with what possible effect may occur in cells, normally expressing large quantities GABP and affected from minor changes in the activated caspase-3 amounts. Notably, cells suiting these requirements are found in some Stem Cell types. Stem Cells are normally expressing much higher amounts of GABP than fibroblast cells, and ESCs, lacking caspase-3, are proven to have ineffective differentiation (Fujita et al., 2008). The authors suggest that the normal differentiation requires caspase-3 – cleavage of Nanog. Although, no connection with the GABP expression level in such cells is established yet, the matching facts of comparatively increased GABP amounts and increased caspase-3 activity in both cell types (our experimental cell cultures and differentiating ESCs) is proposing a link between their mechanisms of action or underlying molecular processes. Supporting the abovementioned results, in caspase-3 – deficient mice is observed accumulation of hematopoietic stem cells with modified differentiation potential (Janzen et al., 2008). These combined data suggest that sufficiently high amounts of GABP might be necessary for the specific "mild" activation of caspase-3 to take effect, possibly via the activation of caspase-12 pathway.

4. SUMMARY

GABP is a heterotetrameric member of *Ets*-family transcription factors. It consists of two subunits – GABP• which contains DNA binding domain and GABP•, which provides transcriptional activation domain and nuclear localization signal. GABP• •• •complex is essential for transcriptional activation of multiple lineage-restricted and housekeeping genes, several viral genes, and in some cases might function as transcriptional repressor. Large variety of data indicates involvement of GABP in the complex regulation of cell growth, specified by quiescence, stimulation/proliferation, apoptosis and senescence. Expression level of GABP• subunit is rapidly increased when resting cells enter S-phase, and GABP• •• •complex is critical to promote the continuity of the cell cycle. Conditional inactivation of GABP• expression in mouse embryonic fibroblasts results in a complete block of proliferation and acquisition of senescence-like phenotype. However, the influence of GABP on the other cell growth determinant – the apoptosis – remains largely obscure. Therefore we aimed to investigate the influence of GABP• •• expression level on the cell growth *in vitro*.

Using siRNA approach we achieved efficient but only transient down-regulation of GABP• expression which precluded further cell growth studies. Persistent increase of the expression of GABP• subunit only resulted in a positive effect on the cell growth speed. Simultaneous conditional overexpression of both GABP• and GABP• subunits though, strongly reduced the growth of the affected cell cultures in reversible and in expression level dependent manner. Interestingly, GABP• •• overexpressing cells did show neither cell cycle arrest nor massive induction of apoptosis. However, more detailed analyses revealed that dampened apoptotic processes were taking place in GABP• ••• overexpressing cells, starting with a prominent activation of caspase-12. Interestingly, activation of downstream effector caspases was rather suppressed explaining a weak increase of apoptotic cells in GABP• •• •overexpressing cultures. This effect suggests that the activation of caspase-12 by elevated amounts of exogenous GABP• •• reflects the normal physiological mechanism of caspase-12 regulation.

ZUSAMMENFASSUNG

GABP ist ein heterotetramerisches Mitglied aus der Familie der *Ets*- Transkriptionsfaktoren. Es besteht aus zwei Untereinheiten – GABP•, welche die DNA-Bindedomäne enthält, sowie GABP•, welche sowohl die Transkriptions-Aktvierungsdomäne als auch das Kernimportsignal umfasst. GABP• •• ist für die Transkriptions-Aktivierung mehrerer differenzierungstypischer als auch sog. *Housekeeping* Gene, sowie einiger viraler Gene essentiell und kann, in einigen Fällen, auch als Transkriptionsrepressor fungieren. Eine Vielzahl von Daten deutet darauf hin, dass GABP in der komplexen Kontrolle des Wachstums von Zellen ein wichtige Rolle zukommt. Dies zeigt sich z. B. im Einfluss von GABP auf zelluläre Vorgänge wie der Stimulation/Proliferation, Apoptose und Seneszenz. So steigt z. B. der Spiegel der GABP• Untereinheit rapide an, nachdem ruhende Zellen die G_0-Phase verlassen und in die S-Phase eintreten. Der aus beiden Untereinheiten gebildete Komplex ist dann für die Progression der Zellen durch den gesamten Zellzyklus von entscheidender Bedeutung. Die Unterdrückung der Expression der GABP• Untereinheit in embryonalen Mausfibroblasten hingegen führt zu einem vollkommenen Proliferations-Stopp dieser Zellen und induziert in diesen einen Seneszenz-artigen Phänotyp. Andererseits ist über den Einfluss von GABP auf andere wichtige das Zellwachstums beeinflussende Faktoren wie z. B. der Apoptose bislang noch recht wenig bekannt. Daher lag es im Fokus dieser vorliegenden Arbeit, den Einfluss der GABP• ••-Spiegels auf das Zellwachstum *in vitro* näher zu untersuchen.

Mithilfe von siRNA-Ansätzen gelang uns die effiziente Herunterregulierung von GABP•. Diese war jedoch nur von vorübergehender Natur, so dass weitere Studien zum Zellwachstum nicht möglich waren. Die stabile Überexpression der GABP• Untereinheit führte dagegen nur zu einem Anstieg der Zellwachstumsgeschwindigkeit. Wurden jedoch sowohl beide Untereinheit gleichzeitig überexprimiert, so resultierte dies in einer deutlichen, Expressionsspiegel-abhängigen und reversiblen Wachstumshemmung der Zellen. Bemerkenswerterweise zeigte die GABP• ••-überexprimierende Zellpopulation weder einen erhöhten Anteil an G_0-Phase noch war eine deutlich ausgeprägte Zunahme der Apoptose-Rate zu verzeichnen. In weiteren Experimenten konnte dennoch eine leichte Erhöhung der Apoptose-Rate in den überexprimierenden Zellen gezeigt werden, was sich durch die deutliche Aktivierung von Caspase-12 belegen ließ. Die Aktivierung von Effektor-Caspasen der Caspase-12 schien allerdings nicht zu erfolgen, was den nur schwach ausgeprägten Charakter der Apoptose zu erklären vermag. Diese Beobachtungen suggerieren, dass die Aktivierung der Caspase-12 durch erhöhte Mengen von exogenem GABP• •• den normalen physiologischen Mechanismus der Caspase-12 Regulation widerspiegelt.

5. MATERIALS AND METHODS

The methods described in this chapter are based on current standard biochemical, molecular and cell biology techniques, often with modifications.

5.1. Materials

5.1.1. Instruments

Hardware	Trade mark
Autoclave	Stiefenhofer
Bacterial shaker	New Brunswick Scientific
Bacterial incubator	Mytron
Balances	Sartorius, Hartenstein
Centrifuges	Heraeus, Beckman
Digital camera - Dimage X1	Konica-Minolta
DNA sequencer 373A	Perkin Elmer
Elisa-reader	Dynatech
Flow cytometer - FACSCalibur•	Becton Dickinson
Gel cameras	Stratagene, Hoefer
Gel dryer	H. Hölzel
Heating block	Hartenstein
Haemocytometer	Paul Marienfeld
Hybridization oven	Bachofer
Ice machine	Genheimer
Intensifying screens	DuPont
Laminar hoods	Heraeus, Gelaire
Light microscopes	Olympus, Leica
Liquid nitrogen tank	Tec-lab
Luminometer	Berthold
Microliter pipettes	Eppendorf, Gilson
Microcentrifuge	Eppendorf
Microcentrifuge (Refrigerating)	Biofuge
Multichannel pipette	Eppendorf
Multi dispencer pipette	Eppendorf

Materials and Methods

PCR machines	Perkin Elmer, MWG
pH meter	Ingold, Hartenstein
Power supplies	Amersham Pharmacia, Biorad
Protein transfer chamber	Hoefer
Quartz cuvettes	Hellma
Refrigerators (–20°C; –70°C)	Privileg, Bosch, Heraeus
RNA/DNA Calculator – GeneQuant	Amersham Pharmacia
Rotors (JA-10, JA-14)	Beckman
Scanner StudioScan II si	AGFA
Scintillation counter	Canberra Packard
Shaker	Hartenstein
SDS-PAGE apparatus	Hoefer, BioRad
Spectrophotometer – GeneQuant Pro	Amersham Biosciences
UV lamp (UVT-20M)	Herolab
Vortexer	Hartenstein
Waterbath	Hartenstein
Water filtration unit (MilliQ Plus)	Millipore

5.1.2. General Materials

Material	Trade Mark
Coverslips	Paul Marienfeld
Cryotubes (2 ml)	Greiner bio-one
Disposable needles, Cuvettes & Syringes	Hartenstein
Glassware	Schott
Micrometer filters (0,2 µM/ 0,45 µM)	Schleicher & Schuell
Nitrocellulose membrane	Schleicher & Schuell
Polypropylene tubes	Greiner bio-one, Nunc
Parafilm	Hartenstein
Pipette tips	Eppendorf
Pipettes	Sarstedt
X-Ray film (13x18 cm, BioMax)	Kodak
Tissue culture plates	Greiner bio-one, Falcon

Materials and Methods

Tissue culture flask (50, 250, 500 ml))	Greiner bio-one
Tissue culture dish (60 mm, 90 mm)	Falcon, Greiner bio-one
Tubes (1,5 & 2 ml)	Sarstedt, Eppendorf
Whatman 3MM paper	Schleicher & Schuell

5.1.3. Chemical Reagents

Reagent	Trade Mark
Acetic Acid [$C_2H_4O_2$]	Carl Roth
Acrylamide/bisacrilamide (29:1) – solution (30%, 40%)	Carl Roth
Agarose	Carl Roth
Ammonium persulfate (APS)	Merck Eurolab
ATP-disodium salt [$C_{10}H_{14}N_5O_{13}P_3Na_2$]	Sigma-Aldrich
•-glycerophosphate [$C_3H_7O_6PNa_2$]	Carl Roth
•-Mercaptoethanol	Carl Roth
Boric Acid	Carl Roth
Bromophenol blue	Merck Eurolab
BSA (Bovine Serum Albumin)	Carl Roth
Butanol [$C_4H_{10}O$]	Carl Roth
Calcium chloride [$CaCl_2$]	Carl Roth
Chloroform [$CHCl_3$]	Carl Roth
Citric acid [$C_6H_8O_7 \cdot H_2O$]	Carl Roth
Coomassie brilliant blue R-250	Roche Applied Science
Crystal violet solution (0,5% (w/v)) plus Methanol (20% (v/v))	Mathilden Apotheke, Würzburg
DAB	Sigma-Aldrich
Disodiumhydrogenphosphate [$Na_2HPO_4 \cdot 7H_2O$]	Merck Eurolab
D-Luciferin [$C_{11}H_8N_2O_3S_2$]	AppliChem
DMEM	Gibco BRL
DMSO	Carl Roth
dNTPs	MBI-Fermentas
DTT	Carl Roth
EDTA [Na_2 EDTA$\cdot 2H_2O$]	Carl Roth

Materials and Methods

EGTA	Sigma-Aldrich
Ethanol [C_2H_5OH]	Carl Roth
Ethidium Bromide [EtBr]	Sigma-Aldrich
FCS	Gibco BRL
Ficoll	Amersham Pharmacia
Formaldehyde [CH_2O]	Carl Roth
Glycerin (87%)	Carl Roth
Glycin [$C_2H_5NO_2$]	Merck Eurolab
Hematoxiline - ChemMate^{TM}	DAKO
Hepes	Carl Roth, Gibco BRL
Hydrochloric Acid [HCl]	Merck Eurolab
Hydrogen Peroxide [H_2O_2]	Carl Roth
Isoamylalcohol	Carl Roth
Isopropanol [2-Propanol, C_3H_8O]	Carl Roth
Leupeptin hydrochloride	Roche Applied Science
L-Glutamine	Gibco BRL
Lithium chloride [LiCl]	Sigma-Aldrich
Milk powder	Saliter
Magnesium acetate [$Mg(C_2H_3O_2)_2 \cdot 4H_2O$]	Sigma-Aldrich
Magnesium chloride [$MgCl_2$]	Carl Roth
Magnesium sulfate [$MgSO_4 \cdot 7H_2O$]	Carl Roth
Manganese chloride [$MnCl_2$]	Fluka
MES [$C_6H_{13}NO_4S$]	Sigma-Aldrich
Methanol [CH_4O]	Carl Roth
Mounting oil (Gelatin - Glycerin)	MERCK
Non-essential amino acids (NEAA, 100x)	Gibco BRL
Phenol [C_6H_6O, TE equilibrated]	Carl Roth
PMSF	Serva
Poly dI/dC	Boehringer Ingelheim
Ponceau Red	Sigma-Aldrich
Potassium acetate [$C_2H_3KO_2$]	Carl Roth
Potassium chloride [KCl]	Sigma-Aldrich
Potassium dihydrogen phosphate [KH_2PO_4]	Sigma-Aldrich
Potassium hydrogen phosphate [$KHPO_4$]	Sigma-Aldrich

Potassium hydroxide [KOH]	Carl Roth
Propidium Iodide (PI 1 mg/ ml ddH$_2$O)	Sigma-Aldrich
Protease inhibitor tablets (complete mini)	Roche Applied Science
Radioactive nucleotides [•^{32}P-ATP, •^{32}P-dCTP, •^{32}P-UTP]	Amersham Pharmacia
Rubidium chloride [RbCl]	Carl Roth
Sodium acetate [CH$_3$COONa•3H$_2$O]	Merck Eurolab
Sodium azide [NaN$_3$]	Merck Eurolab
Sodium bisulfite	Sigma-Aldrich
Sodium carbonate [Na$_2$CO$_3$]	Carl Roth
Sodium chloride [NaCl]	Carl Roth
Sodium fluoride [NaF]	Sigma-Aldrich
Sodium hydrogen phosphate [NaH$_2$PO$_4$•H$_2$O]	Merck Eurolab
Sodium hydroxide [NaOH]	Carl Roth
Sodium orthovanadate [Na$_3$VO$_4$]	Fluka
Sodium pyruvate [C$_3$O$_3$H$_3$Na]	Gibco BRL (100 mM)
Sodium citrate [C$_6$H$_5$Na$_3$O$_7$•2H$_2$O]	Carl Roth
SDS	Carl Roth
Sephadex G50	Amersham Pharmacia
TEMED	Carl Roth
Tris	Carl Roth
Triton X-100	Sigma-Aldrich
Trizol reagent	Gibco BRL
Trypan blue 0,1%	Gibco BRL
Trypsin-EDTA (0,25%)	Gibco BRL
Tween 20	Carl Roth
Xylene cyanol FF	Serva

5.1.4. Solutions and Buffers

All chemicals of molecular biology research grade were purchased from respective manufacturers or suppliers and the solutions were prepared using pure (Milli-Q grade) water. Wherever necessary, solutions were sterile filtered or autoclaved. In the detailed descriptions below only the reagents other than water are listed.

Materials and Methods

APS stock solution (10 %, 10 ml)
    APS .... 1 g

Blocking buffer for Western hybridization
    Fat free milk powder .... 2,5 g
    Dissolved in 50 ml of 1x TBS-Tween

Calcium chloride stock solution (1 M, 1000 ml)
    $CaCl_2$ (anhydrous).. 110,98 g

Colony hybridization solutions
    <u>CH solution I (freshly prepared)</u>
    NaOH .... 0,5 N
    <u>CH solution II</u>
    Tris- HCl (pH 7,5) .... 0,2 M
    NaCl .... 1,0 M
    <u>CH solution III</u>
    SDS (Sodium Dodecyl Sulfate)................................. 1%
    EDTA.. 1 mM
    Na_2HPO_4 (pH 6,8) .... 40 mM
    <u>CH pre-hybridization solution</u>
    SSC (Saline-Sodium Citrate)..................................... 2x
    SDS 0,2%
    Denhardt solution 1x
    Salmon sperm DNA .... 100 • g/ml
    <u>CH hybridization solution</u>
    SSC 5x
    SDS 0,2%
    Denhardt solution 2x
    Salmon sperm DNA .... 100 • g/ml
    <u>CH washing buffer</u>
    SSC 2x
    SDS 0,2%

Coomassie blue solution (1000 ml)
    Coomassie Brilliant Blue R-250 .... 2,5 g
    Methanol .... 450 ml
    Acetic Acid .... 100 ml

DAB - 3'3'tetra-diamino-benzidine (1x working concentration)
   979 ml Phosphate-Buffered Saline (PBS)
   20µl DAB (0,03g/ml stock solution)
   1µl H_2O_2

Denaturing PAA-Gel for RNase Protection Assay (1000 ml; 6% gel solution)
   Urea ....280,0 g (8M)
   30% Acryl-Bisacrylamide solution .... 240 ml
   10x TBE buffer ....100 ml
   Composition for one gel (Polymerization takes approx. 1-2 hours at RT)
   6% gel solution .... 30 ml
   10% APS .... 300 µl
   TEMED (Tetramethylethylenediamine)....................... 60 µl

DNA Electrophoresis Buffer (1000 ml)
   TAE (50x) .... 20 ml

DNA gel composition:

	0,7%	1,0 %	2,0 %
Agarose	1,05 g	1,5 g	3,0 g
20x TAE	7,5 ml	7,5 ml	7,5 ml
ddH_2O	142,5 ml	142,5 ml	142,5 ml
EtBr (5mg/ml)	25 µl	25 µl	25 µl

DTT stock solution (1 M, 20 ml)
   DTT .... 3,09 g
   DTT powder was dissolved in water, sterilized by filtration (must not be autoclaved), aliquot in Eppendorf microcentrifuge tubes and frozen at –20°C.

EDTA stock solution (0,5 M, 1000 ml)
   Na_2 EDTA·$2H_2O$ ....186,1 g
   pH of the solution was adjusted to 8,0 with 10 M NaOH (~ 50 ml)

EGTA stock solution (0,25 M, 1000 ml)
   EGTA .... 95 g
   pH of the solution was adjusted to 8,0 with KOH

EMSA Solutions

PAA gel for the purification of radioactive DNA probes (50 ml, 12%)
ddH$_2$O	30 ml
30% acryl-bisacrylamide solution	15 ml
10x TBE buffer	5 ml
10% APS	300 µl
TEMED	60 µl

EMSA gel (6%, 100 ml)
dd H$_2$O	70 ml
30% acryl-bisacrylamide solution	20 ml
10x TBE buffer – (till 0,4xTBE final)	4 ml
10% APS	500 µl
TEMED	50 µl

EMSA binding buffer (3x, 50 ml)
1M Hepes/KOH (pH 7,9)	3 ml
1 M KCl	7,5 ml
0,5 M Na$_2$EDTA·2H$_2$O (pH 8,0)	300 µl
1 M DTT	150 µl
Ficoll	6 g

Aliquotes were stored at –20°C.

EMSA running buffer – 0,4x TBE (1x, 1000 ml)
TBE (10x)	40 ml
H$_2$O (Milli-Q grade)	960 ml

Ethidium bromide stock solution (100 ml)

EtBr	1 g

The solution was stored at 4°C in a dark bottle.

FACS buffer (stored at 4°C)

10x PBS, pH 7,4	50 ml
Cell culture grade H$_2$0	450 ml
1,0 M Sodium azide	0,5 ml

The solution was sterile filtered and stored at 4°C.

Gel loading sample buffer, 6x (MBI Fermentas)

Glycerin	60%
EDTA	60 mM
Bromophenol blue	0,09%

Xylene Cyanol FF 0,09%

2x Hepes-Buffered Saline (HBS)
- Hepes/KOH (pH 7,05) .... 50 mM
- KCl .... 10 mM
- Dextrose .... 12 mM
- NaCl .... 280 mM
- Na_2HPO_4 .... 1,5 mM

The solution was sterile filtered through 0,2 µm filter, aliquot and stored at −20°C.

HEPES/ KOH stock solution (1 M, 1000 ml)
- HEPES 238,33g

pH of the solution was adjusted to 7,2/ 7,4/ 7,9 with KOH.

Luciferase Harvesting Buffer (50 ml)
- 1,5 M Tris/HCl (pH 7,8) .... 1,7 ml
- 1 M MES (2-(N-morpholino) Ethane Sulfonic Acid)..... 2,5 ml
- Triton X-100 .... 50 µl

The solution was freshly prepared and 50 µl DTT stock solution (1M) was added just before use.

Luciferase Assay Buffer (50 ml)
- 1,5 M Tris/HCl (pH 7,8) .... 4,17 ml
- 1 M MES .... 6,25 ml
- 1 M Mg $(C_2H_3O_2)_2 \cdot 4H_2O$ .... 1,25 ml

The solution was freshly prepared and supplemented with ATP just before use.

Luciferin solution for Luciferase Assay (100 ml)
- Luciferin .... 28 mg
- 1 M $KHPO_4$ (pH 7,8) .... 0,5 ml

The solution was aliquot and stored at −20°C.

Nuclear and cytoplasmic extract preparation buffers

<u>Extraction buffer A (Hypotonic, 1000 ml)</u>
- 1 M Hepes/KOH (pH 7,9) .... 10 ml
- 1 M KCl .... 10 ml
- 0,5 M $Na_2EDTA \cdot 2H_2O$ (pH 8,0) .... 200 µl
- 0,25 M EGTA (pH 8,0) .... 400 µl

Solution was stored at 4°C. For preparing buffer A⁺, the following inhibitors were added before experiment: DTT till 0,1 mM (stock solution 1 M or 0,5 M) and PMSF till 2 mM.

Extraction buffer C (High salt, 1000 ml)

1 M Hepes/KOH (pH 7,9)	20 ml
1 M KCl	400 ml
0,5 M $Na_2EDTA \cdot 2H_2O$ (pH 8,0)	2 ml
0,25 M EGTA (pH 8,0)	4 ml

Solution was stored at 4°C. To prepare buffer C⁺, the following inhibitors were added before experiment: DTT till 1 mM (stock solution 1 M or 0,5 M) and PEFAblock till 2 mM.

PBS (10x, 1000 ml)

NaCl	80 g
KCl	2 g
$CaCl_2$	1 g
$MgCl_2$	1 g
$Na_2HPO_4 \cdot 7H_2O$	26,8 g
KH_2PO_4	2,4 g

pH of the solution was adjusted to 7,4 with 1 N HCl.

Phosphatase inhibitor (stock solution, final working concentration is indicated)

Sodium Orthovanadate [Na_3VO_4 stock solution (0,2 M): 4 mg/ml in H_2O]:1 mM

pH of the solution was adjusted to 10,0 with 1 N NaOH or 1 N HCl (solution becomes yellow), boiled for 10 min at 100°C (solution becomes colourless), cooled to RT and subsequently pH was again adjusted to 10,0; this was repeated till solution becomes colourless at RT and pH gets stabilized at 10,0. Aliquots were stored at –20°C and just before use boiled for 5 min at 100°C and left at RT to cool down.

Ponceau red solution (100 ml)

Ponceau red	2,0 g

Potassium chloride stock solution (1 M, 1000 ml)

KCl	74,6 g

Potassium hydrogen phosphate stock solution (1 M, 1000 ml)

$KHPO_4$	135,1 g

pH of the solution was adjusted to 7,8 with KOH.

Potassium phosphate buffer (0,2 M, 1000 ml)

    KH_2PO_4 ....27,2 g

    pH was adjusted to 7,0 with 1 M KOH.

Propidium Iodide Solution

    10μg/ml Propidium Iodide (Sigma, Cat#P4170, MW: 668.4) in PBS (pH 7.4)

Protease inhibitors (final working concentration is indicated)

    Aprotinin   [Stock solution (0,3 M): 2 mg/ml H2O]   0,3 μM

    Leupeptin   [Stock solution (2 mM): 1 mg/ml H2O]   2 μM

    AEBSF   [Stock solution (0,2 M): 50 mg/ml H2O]   1 mM

    Protease inhibitors cocktail tablets [1 tablet/50ml buffer]

RIPA buffer (3x, stored at RT)

    1% Tryton X-100

    0,3% NaDOC (Deoxycholic Acid Sodium salt)

    0,3% SDS

    0,42M NaCl

    0,15M Tris (pH 7,5)

    3mM EDTA/EGTA

    15mM NaF

    0,002% NaN_3

RNase Preparation

    10mM Tris + 15mM NaCl + RNase to final dilution of 10mg/ml. Boiled for 5 min for DNase deactivation, aliquot and stored frozen at –20°C.

SDS stock solution (10%, 1000 ml)

    SDS .... 100 g

SDS-PAGE sample buffer (4 x, 100 ml)

    1,5 M Tris/HCl (pH 6,8) .... 20 ml

    SDS ....2,4 g

    Glycerine (87 %) ....50 ml

    •-Mercaptoethanol .... 25 ml

    Bromophenol blue 0,04%

    The solution was warmed to 70°C and stored at –20°C.

SDS-PAGE separating (lower) buffer

    1,5 M Tris (pH – 9,0)

    0,4% TEMED

0,4% SDS

SDS-PAGE stacking (upper) buffer
- 0,14 M Tris (pH – 6,8)
- 0,11% TEMED
- 0,11% SDS

SDS-PAGE running buffer (10x, 1000 ml)
- Tris .... 30,3 g
- Glycin .... 144,1 g
- 10% SDS .... 100 ml
- pH of the solution was adjusted to 8,5 with 1 N HCl.

Sodium acetate stock solution (3 M, 1000 ml)
- $CH_3COONa \cdot 3H_2O$ .... 408,24 g
- pH of the solution was adjusted to 5,2 with concentrated acetic acid.

Sodium chloride stock solution (5 M, 1000 ml)
- NaCl .... 292,22 g
- The solution was dissolved by heating to 60°C.

Sodium hydroxide stock solution (10 M, 1000 ml)
- NaOH .... 400 g

Sodium phosphate buffer (0,2 M, 1000 ml)
- $NaH_2PO_4 \cdot H_2O$ .... 27,6 g

Stripping buffer for nitrocellulose membrane (1000 ml)
- 1,5 M Tris/HCl (pH 6,8) .... 41,7 ml
- 10% SDS .... 200 ml
- Before use, 100 ml of buffer was supplemented with 700 µl •-Mercaptoethanol.

TAE buffer (Tris/Acetate/EDTA, 50x, 1000 ml)
- Tris .... 242 g
- 0,5 M $Na_2EDTA \cdot 2H_2O$ (pH 8,0) .... 100 ml
- Concentrated acetic acid .... 57,1 ml

TBE buffer (Tris/Borate/EDTA, 10x, 1000 ml)
- Tris .... 108 g
- Boric Acid .... 55 g
- 0,5 M $Na_2EDTA \cdot 2H_2O$ (pH 8,0) .... 40 ml

TBS (Tris-Buffered Saline, 20x, 1000 ml)
- Tris .... 121 g

NaCl .... 175,2 g

KCl .... 7,5 g

    pH of the solution was adjusted to 7,6 with 1M HCl (~ 10,2 ml)

TBS/Tween (TBS-T, 1x, 1000 ml)

    TBS (20x) .... 50 ml

    Tween 20 .... 1 ml

TE buffer (Tris/EDTA, pH 8,0, 1000 ml)

    1,5 M Tris/HCl (pH 8,0) .... 6,7 ml

    0,5 M EDTA (pH 8,0) .... 0,2 ml

Transfer buffer for Western blot (1000 ml)

    Glycine .... 2,9 g

    Tris .... 5,8 g

    10% SDS .... 3,7 ml

    Methanol .... 200 ml

Tris/ HCl stock solution (1,5 M, 1000 ml)

    Tris .... 181,7 g

    pH of the solution was adjusted to 6,8/ 7,5/ 7,8/ 8,0/ 8,8 with 1 N HCl.

Whole cell extract preparation buffer (Kyriakis lysis buffer modified, 1000 ml)

    1 M Hepes/KOH (pH 7,4) .... 20 ml

    0,25 M EGTA (pH 8,0) .... 8 ml

    NaF .... 2,1 g

    •-Glycerophosphate .... 10,8 g

    Glycerine (87%) .... 115 ml

    Triton X-100 .... 10 ml

    NaN_3-solution (10%) .... 4 ml

The solution was stored at 4°C. For 10 ml KLBM$^+$ buffer, the following inhibitors were added before the experiment: 10 µl DTT-stock solution (1M), protease inhibitors [50 µl AEBSF (0,2 M), 10 µl leupeptin (2 mM) and 10 µl aprotinin (0,3 M)] and phosphatase inhibitor [50 µl Sodium orthovanadate (0,2 M)].

5.1.5. Antibiotics

Ampicillin	Hoechst
Doxycycline	Sigma-Aldrich
Neomycin (G-418)	Invitrogen

Materials and Methods

Penicillin (10,000 IU/ml)	Hoechst
Puromycin	Sigma
Streptomycinsulphate (10 mg/ ml)	Hoechst
Ciprofloxacin	Mediatech

5.1.6. Kits

Agarose gel extraction kit (QIAEX II)	QIAGEN
BioRad protein assay (5x Bradford reagent)	BioRad
ECL Chemiluminescence Kit	Amersham
Gel extraction kit (Jetsorb)	Genomed
PCR Purification Kit (QIAquick)	QIAGEN
Plasmid DNA Isolation kit (Midi)	Macherey-Nagel
Plasmid DNA Isolation kit (Mini)	QIAGEN
RNase Protection Assay Kit (RiboQuant)	BD Pharmingen
TaqDyePrimer sequencing Kit	Perkin Elmer
Transfection reagents (SuperFect• , PolyFect•)	Qiagen
Western blotting substrate (Lumi light)	Roche

5.1.7. DNA Size Markers

The GeneRuler 100 bp and 1 kb DNA size markers were purchased from MBI-Fermentas. The size of fragments in markers was as following:

100 bp Marker   1031 / 900 / 800 / 700 / 600 / 500 / 400 / 300 / 200 / 100 / 80 [bp]
1 kb Marker   10000 / 8000 / 6000 / 5000 / 4000 / 3500 / 3000 / 2500 / 2000 / 1500 / 1000 / 750 / 500 / 250 [bp]

5.1.8. Protein Standards

The protein size marker BENCHMARK• was obtained from Gibco BRL.

5.1.9. Enzymes

Restriction endonucleases and modifying enzymes	MBI-Fermentas,

SAWADY PWO DNA polymerase	New England Biolabs
peqGOLD Pwo-DNA-Polymerase	Peqlab
Proteinase K and RNase [Ribonuclease] Type I-A	PEQLAB
Streptavidin-HRP - (LSAB 2 System)	Sigma-Aldrich
	DAKO Cytomation

5.1.10. Antibodies

5.1.10.1. Primary

Ag (mouse)	Ab	Format	Clone	Manufacturer
GABP•	Rabbit	Purified	N/A	Kindly provided by Prof. U. R, Rapp, Uni. Würzburg
GABP•.	Rabbit	Purified	N/A	Kindly provided by Prof. U. R, Rapp, Uni. Würzburg
•-Actin	Mouse	Purified	AC-15	Sigma-Aldrich
PARP	Rabbit	All forms	#9542	Cell Signaling
Caspase-12	Rabbit	polyclonal	(M-108)	Santa Cruz

The antibodies against the caspases -3, -7, -9 and -12 were kindly provided by Dr. Ingolf Berberich, Uni. Würzburg.

5.1.10.2. Secondary

Goat anti-rabbit IgG Peroxidase conjugate - Sigma, Polyclonal
Goat anti-mouse IgG Peroxidase conjugate - Sigma, Polyclonal
Biotinylated secondary anti-rabbit + anti-mouse antibodies (solution 1) - DACO Cytomation LSAB2 system HRP

5.1.11. Oligonucleotides and primers

All oligonucleotides were produced by MWG Biotech, dissolved in water (100 pmol/μl) and stored at –80°C or –20°C (long-term storage) or at +8°C (short-term).

5.1.11.1. Oligonucleotides for the Construction of siRNA-expressing Primary Vectors

Materials and Methods

Only the oligonucleotides used for creating of siRNA constructs, giving the best results are shown here. The full list of siRNA sequences used for the screening is presented in the Appendix "Design of RNAi targets".

Plasmid	Position in gene seq.	Oligo's sequence
pGAsiR 10	107	5`-gatccccTGTGTAAGCCAGGCCATAGttcaagagaCTATGGCCTGGCTT ACACAtttttggaaa-3`
pGAsiR 7	74	5`-gatccccAGCATTGTGGAACAAACCTttcaagagaAGGTTTGTTCCACA ATGCT tttttggaaa-3`

5.1.11.2. Olygonucleotide for EMSA

DSE$_{TSC2}$ 5'- TCCGCTA**CCGGAAG**TGCGGGTCGCG**CTTCCGG**CGGCGT -3'
3'- AGGCGAT**GGCCTTC**ACGCCCAGCGC**GAAGGCC**GCCGCA -5'

5.1.11.3. PCR Primers for the Amplification of Inserts Used for DNA Cloning

Plasmid		Primer name	Primer sequence [5' - 3']
pTRE-GABP·	F	mGAa5'BglIIpure	TCA **GAG ATC T**GC ACC ATC ACT AAG AGA GAAGCA GAA G
	R	mGAa-3' XbaI	TCA **GTC TAG A**TC AAA TCT CTT TGT CTG CCT G
pTRE-GABP· and	F	mGab1-5'-BglIIpure	TCA **GAG ATC T**GC ACC ATG TCC CTG GTA GAT TTG GG
pMSCVpG-GABP·	R	GABPb-HpaI	TCA **GGTTAAC**TT ATT TTG GAT GGC TGC AGC A
pMSCVpG and	F	IRES-GFP F2-ClaI	TAT AT**A TCG AT**C GCC CCT CTC CCT CCC C
pRS10-GFP	R	IRES-GFP R-ClaI	ATA TAT **CGA T**GC TTT ACT TGT ACA GCT CGT C

5.1.12. Plasmid Constructs

pSUPER   kindly provided by Prof. Dr. Reuven Agami
psiGA#7   constructed in the span of this work
psiGA#10   constructed in the span of this work
pMSCVpuro   Clontech
pMSCVpuro-·   constructed in the span of this work
pRS10   constructed in the span of this work
pMSCVpuroG   constructed in the span of this work
pRS10G   constructed in the span of this work

pMSCVpG-GABP•	constructed in the span of this work
pTRE-6xHN	Clontech
pTRE-GABP•	constructed in the span of this work
pTRE-GABP•	constructed in the span of this work
pMSCVpuro-• -TRE-GABP•	constructed in the span of this work
pMSCVpuro-• -TRE-GABP• ••	constructed in the span of this work
pTet-ON	Clontech
pTet-OFF	Clontech
pMSCVneo	Clontech
pTet-ON-S2	kindly provided by Dr. Hermann Bujard
pMAX-EGFP	AMAXA
pLX	kindly provided by Dr. Wilson A.C.
pLX-GABP•	kindly provided by Dr. Wilson A.C.
pLX-GABP•	kindly provided by Dr. Wilson A.C.

For detailed clone charts of the plasmids, created during this work, please see the Appendix "Clone charts".

5.1.13. Growth Media

5.1.13.1. Mammalian Cell Culture Medium for Adherent Cell Lines

Preparation: The listed compounds are mixed together under sterile conditions (in Laminar Flow Hood)

DMEM	500 ml
FCS	10%
L-Glutamine (200 mM)	5 ml
Penicillin (10,000 IU/ ml), Streptomycinsulphate (10 mg/ ml) - 3 ml	
(Or Ciprofloxacin	till 25 µg/ml)
•-Mercaptoethanol (50 mM)	500 µl
Sodium pyruvate (100 mM)	5 ml
Non-essential aminoacids (NEAA, 100x)	5ml

5.1.13.2. Bacterial Culture Media

LB broth, Lennox	Difco

LB agar (LB broth/1,5% Agar-Agar, Carl Roth)
Terrific Broth (TB) - powder   Sigma-Aldrich
4ml/L Glycerol were added prior preparation

5.1.14. Mammalian Cell Lines
HIN-3T3	kindly provided by Dr. Jakob Troppmair
293T	Genome Systems Ltd
U2OS	kindly provided by Prof. Dr. Stephan Ludwig

5.1.15. Bacterial Strains
E. coli strains "JM109", „DH5•", „XL-1 blue", „SURE" (Stratagene) and "Top10" (Invitrogene) were used.

5.1.16. Software
Microsoft Office (Word, Excel and Power Point) was used for routine tasks of data management. For image processing *Adobe Photoshop* and *Corel Draw Graphic Suite* were used. X-ray films (Western blots, EMSAs, etc.) were scanned using *FotoLook SA (2.08)*. DNA sequence analyses were performed with *NTI Vector* and several available online tools, flow cytometry data were analysed using *Cell Quest Pro*. Densitometry was performed with *FotoLook SA (2.08)* and *ImageJ 1.45p*. The list of citations used in this work was created and managed with *EndNote*. The program *Viper* was used for keeping this work free of "plagiarism, missing quotation marks and incorrect citations".

5.2. Methodology

5.2.1. DNA Methods
5.2.1.1 Isolation of Plasmid DNA
5.2.1.1.1 Small Scale Plasmid DNA Purification
Single bacterial colonies were inoculated in 2,5ml LB^{Amp} medium and cultivated with shaking overnight at 37°C. Next day, the bacteria were collected by centrifugation and the plasmid DNA isolated using the QIAprep Spin Miniprep Kit (QIAGEN), according to the instructions of the manufacturer. ~500µl of the bacterial culture was left at 4°C for inoculation of maxi preps.

5.2.1.1.2 Medium Scale Plasmid DNA Purification

250 ml of TB^{Amp} medium was inoculated with 50-100µl of bacterial culture, used for mini prep and cultivated with shaking overnight at 37°C. Next day, the bacteria were collected by centrifugation and the plasmid DNA isolated using the NucleoBond® Midi/Maxi prep kit (Macherey-Nagel), according to the instructions of the manufacturer.

5.2.1.2 Determination of DNA/RNA Concentration

DNA/RNA concentration was determined by UV spectrophotometeric measurements.
A quartz cuvette was washed twice with distilled water and filled with 100µl of ddH$_2$O. Empty reference was taken on spectrophotometer (OD260) and the water was replaced with the sample (2µl of DNA/RNA solution diluted in 98µl ddH$_2$O). After OD260 sample measurement the concentration of DNA/RNA solution was calculated automatically (dilution factor 50).

5.2.1.3 DNA Electrophoresis on Agarose Gel

See chapter 5.1.4 for DNA gel composition.

The suspension was boiled in the microwave oven until the agarose was completely dissolved. Then the solution was cooled down to around 50•C and ethidium bromide (EtBr) was added up to 0,5 •g/ml. The gel was poured to solidify into the gel casting chamber. Appropriate comb was used for forming the slots and all was cooled down to RT. DNA gel loading buffer was added to the DNA samples and they were applied into the slots. The gel was run in 1x TAE buffer at 75-150 mA for a desired duration. The DNA was visualised under UV-light and picture was recorded in digital form or printed out.

5.2.1.4 Isolation of DNA from Agarose Gel

The gel slices containing the desired DNA fragments were excised from preparative agarose gel under UV light at low settings of the transilluminator, exposing on UV light for as short time as possible and the DNA was extracted according to the instructions of manufacturers (QIAEX II kit, QIAGEN and Jetsorb, Genomed).

5.2.1.5 Restriction Enzyme Digestions of DNA

2-4µg DNA (optimal concentration in the reaction mix 0,1 – 0,2 µg/µl) were mixed with 2µl 10x digestion buffer, 2µl restriction enzyme (10U/µl) and ddH$_2$O to end volume of 20 µl.

The digestion reaction was incubated at 37°C for 1 hour to overnight. The completeness of the reaction was resolved on an agarose gel.

5.2.1.6 Polymerase Chain Reaction (PCR)

The polymerase chain reactions included a number of variables, slightly differing under the differential conditions of primer sets and DNA templates. One example of polymerase chain reaction conditions is presented below:

25 µl PCR reaction containing 1x PCR buffer, 250 µM dNTP mix, 1-4 mM $MgCl_2$, 60 pmol from each primer, 0,25-0,5 U Taq DNA polymerase and 10-100 ng DNA template was incubated in the thermal cycler at 95°C for 2-5 min for DNA chain separation and then ran 30-40 cycles at the following program:

95°C for 15-25 seconds – denaturing step

T_A°C for 40 seconds – annealing step

72°C for 40-60 seconds – extension step

Then the reaction was held at 72°C for 7 min and placed at 4°C for short term storage.
The PCR products were resolved on an agarose gel and visualized under UV light.

5.2.1.7 Purification of DNA

DNA products, resulting from PCR and other enzymatic reactions were purified by QIAquick PCR Purification Kit, according to the instructions of the manufacturer (QIAGEN) and using the supplied protocol.

DNA products designated for sequencing were purified on self-made Sephadex G-50 columns as follows:

100µl filter tips were inserted through the caps of 1,5ml screw-capped tubes. The Sephadex suspension was filled in the tips and shortly span at 3000 rpm. The cap/column assemblies were transferred on new tubes with 1/10 sample volume Na Acetate on the bottom. The amplification product was applied on the Sephadex and span for 3 min at 3000 rpm. The DNA was precipitated by addition of 5 volumes Ethanol. The yield was harvested by spinning at max speed for 5 min, aspirating the supernatant and drying at 60°C for 5 min.

5.2.1.8 Ligation of DNA Fragments

The desired plasmid DNA was digested with the desired restriction enzyme, dephosphorylated by treatment with alkaline phosphatase and purified. 100 ng plasmid DNA and appropriate amounts of insert DNA (in two series in molar ratios 1:10 and

1:100) were combined with corresponding amounts ddH$_2$O, ligation buffer and 2 μl T4 DNA ligase (2 Weiss units) in a total volume of 30 μl per reaction. The reaction was incubated at 16°C for 4 hours or overnight. The ligation product was transformed into competent bacterial cells.

5.2.1.9 Colony Hybridization

Nitrocellulose membrane was cut in circular shape with diameter 0,5 mm less than the diameter of the bacterial plate and placed over the bacterial colonies for 2-3 min. The membrane was labelled and abruptly removed. The cells were lysed by placing the membrane for 5-10 min on blotting paper, soaked with CH solution I, followed by placing twice for 10 min on blotting paper with CH solution II. Then the membrane was air-dried, soaked with CH solution III for 10 min and dried again, followed by baking for 30 min at 80°C in hybridization oven. The pre-hybridization was done in 15-20 ml of CH pre-hybridization solution for 30 min at temperature 3-4 °C lower than the T$_A$ of the probe oligo. The hybridization was in 15-20 ml of CH hybridization solution, supplemented with the labelled probe for 1h at the same temperature. The membrane was washed in 300-400 ml of CH washing buffer, air-dried and exposed on x-ray film overnight.

5.2.1.10 PI Staining and Flow Cytometry

The cells of interest were harvested by trypsinization, span down in FACS tubes, resuspended in 0,25ml BSA-containing PBS and cooled on ice. Without any delay 0,75ml of cold 100% Ethanol was added dropwise with continuous swirling of the tube. The samples were fixed by keeping them for minimum 10 min at –20°C. Four hours prior to FACS analysis the fixed cells were mixed with 4ml PBS by inversion of the tubes several times, centrifuged at 400x g for 5min, resuspended in 250μl propidium iodide solution, supplemented with 5μl of 10mg/ml RNase (MBI Fermentas, DNase free) and incubated at 37°C for 4h.

The Flow Cytometry for all types of measurements (PI, GFP and size/granularity) was done on flow cytometer - FACSCalibur• where the settings were adjusted according to the requirements of every particular experiment.

5.2.2 RNA Methods

5.2.2.2 RNA Isolation from Mammalian Cells

Total cellular RNA was isolated using TRIZOL reagent from up to 1×10^7 cells/ml reagent and the procedure was followed according to the instructions of manufacturer.

5.2.2.1 Ribonuclease Protection Assay

RNase protection assays were used for quantitative determination of mRNA. The RNase Protection Assay Kit (RiboQuant) from BD Pharmingen was utilized and the procedure was followed according to the instructions of manufacturer. The template sets mCyc1 and mCyc2) were employed to quantitatively determine the transcription levels of the mouse cyclins: A1, A2, B1, B2, C, D1, D2, D3, E, F, G1, G2, H and I.

5.2.3 Protein Methods

5.2.3.1 Preparation of Protein Extracts

Harvesting adherent mammalian cell cultures:
Cells were detached by treating with trypsin for 5 min at 37°C, mixed with 1 vol. FCS-containing growth media, centrifuged (1400 rpm, 4 min, 4°C), washed with cold PBS (without Ca^{++} and Mg^{++}) and again centrifuged at 8000 rpm for 30 sec at RT.

When the purpose is preparation of nuclear and cytoplasmatic protein extracts, an alternative way of harvesting is applied: The media is aspirated and cells are washed once with cold PBS (Without Ca^{++} and Mg^{++}), the bottom of the tissue culture dish is covered with buffer A^+ and incubated 10 min on ice, followed by stripping the cells with vial or rubber policeman. Then the pellet is span down at 3500 rpm for 1 min.

5.2.3.1.1 Preparation of Whole Cell Protein Extracts (for Western Blot)

The cells were resuspended in RIPA buffer supplemented with proteinase inhibitors, vortexed briefly and frozen at –70°C. When needed the samples were thawed on ice, vortexed at 4°C for 15 min, cleared by centrifugation at max speed for 10 min at 4°C and supernatant transferred into new tubes. The protein concentration of the extracts was determined by Bradford protein assay (Bio-Rad), see below. (Bradford, 1976)

5.2.3.1.2 Preparation of Nuclear and Cytoplasmatic Protein Extracts (for EMSA)

Materials and Methods

The fresh cell pellet was resuspended in 200 µl to 1 ml (depending on the pellet volume) of Extraction Buffer A⁺ (100 µl per 1×10^7 cells) and incubated for 10-20 min on ice. The solution was passed 10 times through 1 ml syringe with 26G needle and centrifuged at 8000 rpm for 1 min at 4°C. The supernatant contains cytosolic fraction and the pellet, which appear transparent, contains nuclear fraction. The supernatant was transferred to a new tube and kept on ice. The pellet was washed with 800 µl extraction buffer A⁺, centrifuged at 8000 rpm for 1 min at 4°C and resuspended in extraction buffer C⁺ (leupeptin was added in addition to DTT and PMSF), followed by vortexing the nuclear extract vigorously for 30 minutes at 4°C. The suspension was centrifuged at 14000 rpm for 30 minutes at 4°C and supernatant, containing nuclear proteins was frozen at –70°C. The protein concentration of the extracts was determined by Bio-Rad protein assay.

5.2.3.2 Determination of Protein Concentration

Equal volumes (2µl) of cell lysate containing protein was added to 998 µl of diluted dye reagent (1:5 dilution of dye reagent concentrate in ddH$_2$O), mixed well, incubated 15 min at RT for colour stabilization and O.D. value was measured at 595 nm. The corresponding extraction buffer was always included in the blank control as also 2 µl of buffer + 998 µl of Bradford reagent. The real protein concentration was determined by comparing the O.D. value of the sample against standard curve for BSA and recalculating for the dilution (Bradford, 1976; Reisner, 1975).

5.2.3.3 SDS-PAGE and Immunodetection

(In the other chapters of the thesis to SDS-PAGE together with Western blotting and hybridization is referred as Western blot analyses)

5.2.3.3.1 SDS-Polyacrylamide Gel Preparation and Electrophoresis

SDS-polyacrylamide gels were prepared in 8 cm x 10 cm x 0,75 mm mini gel format according to the standard Laemmli method (Laemmli, 1970).

For separation of proteins 10% or 12% SDS-PAAGs were used for proteins in the range of 20-300 kDa and 10-200 kDa respectively.

Separating or lower gel mix was prepared according to the volume required, poured in the gel apparatus, overlaid gently with 75 % Ethanol and left to polymerize at 37°C for 5 minutes.

After the separating gel was polymerized, the overlay was decanted and gently washed with distilled water. The stacking gel was poured, the comb was inserted and allowed to polymerize for 5 min at 37°C. Required concentration of protein samples were mixed with 4x Laemmli buffer and denatured by heating at 95°C for 5 min, loaded in the wells of polymerized gel and run at constant current, 35-45 mA per gel, in 1x SDS-PAGE running buffer until the front reaches the end of the gel.

Table 5.1: Composition of protein gels (all numerical figures are in ml)

Gel percentage:	10% Separating gel	12% Separating gel	4% Stacking gel
Acryl-/ Bisacrylamide			
- 30% →	3,35	4	1,3
- 40% →	2,5	3	1
Buffer	Lower	Lower	Upper
(A/B-30%) →	2,5	2,5	8,6
(A/B-40%) →	2,5	2,5	8,9
H₂O			
(A/B-30%) →	4,05	3,4	none
(A/B-40%) →	4,9	4,4	none
APS - 10%	0,1	0,1	0,1
Total Volume	10	10	10

5.2.3.3.2 Western Blotting and Hybridization

The SDS-PAGE gel was electro transferred onto nitrocellulose membrane at 200 mA for 1,5-2h on ice. The membrane was incubated (from now on with gentle rocking) in a blocking solution (5% fat free milk in 1x TBS-Tween) for 1 hour at RT followed by incubation with primary antibody solution (1:1000 in blocking solution + 0,04-0,08% NaN_3) for 1 hour at RT. Membrane was rinsed and washed in 1x TBS-Tween for 4x 5 minutes followed by incubation with secondary antibody-HRP conjugate solution (1:10000 in blocking solution) for 45 min at room temperature, rinsed and washed in 1x TBS-Tween for 1x 15 minutes and 5x 5 minutes. Detection was performed with ECL developing solution (Amersham) or Western blotting substrate – LumiLight (Roche), according to the instructions of the manufacturer.

Materials and Methods

5.2.3.3.3 Stripping of Nitrocellulose Membrane

To remove the bound antibodies membrane was incubated in stripping buffer for 30 min at 60°C with shaking, washed once with ddH$_2$O, three times with TBS/Tween for 5 min each and after additional blocking step used for blotting with another primary antibody.

5.2.3.3.4 Cytostaining of Adherent Cells (NIH-3T3)

Microscope glass slides were sterilized by incubating in 75% ethanol for 5-10 min, washed once with sterile PBS and submerged under a layer of pre-warmed growth media in Petri dishes. The required cells were trypsinized, mixed with 1 volume media and plated at appropriate densities over the slides. The culture vessels were left in strictly horizontal position for 1 hour until the cells attach and than incubated in tissue culture incubator overnight.

Next morning the plates were removed from the media, washed once with cold PBS and dipped for 10 min in -20°C 100% acetone for cell fixation (incubation is done in freezer). Then they were air dried for 30 min at RT and washed 3 times, 5 min each in cold PBS. The excess cells from the sides of the slides were removed with paper towel, forming the staining areas. (All further incubations of solutions on the glass plates were done in humid chamber, preventing evaporation of the small amounts of used solutions). The staining areas were covered with 100µl of primary antibody, diluted in PBS 1:1000 and incubated minimum 1h at RT. The antibody solution was removed and plates washed 3x5 min with cold PBS. The primary antibody was detected by DACO LSAB2 System, containing biotinylated secondary anti-rabbit + anti-mouse antibodies (solution 1) and streptavidin-HRP (solution 2). Each from the solutions was incubated on the stained areas for 10 min at RT, followed by washing with cold PBS 3x5 min. The signal was revealed by incubation with 100µl 1x working solution of DAB/H$_2$O$_2$ for maximum of 10-15 min for obtaining a brown colour of the desired protein. At the end the slides were washed with distilled H$_2$O.
(After this step the slides can be sealed with cover slips)

For visualization of the cellular nuclei or other DNA-containing structures the slides were incubated in Meyer's hematoxiline solution (red colour) for 1 min and then washed under tap water for 1-2 min until blue colour appears. For stopping the increase in the colour intensity, slides were washed with distilled water and dried.

For mounting the cover slips, a drop of pre-warmed mounting oil was layed over the stained area and the cover slip was sealed by levelling it from aside, thus expelling the air bubbles.

5.2.3.4 Luciferase Assay

The experimental cells were harvested by trypsinization and washed with PBS. The centrifuging steps were done in microcentrifuge at 8000 rpm. After the harvesting step the procedures were done on ice or at 4°C.

75-100μl (depending on the cell pellet volume) of Harvesting Buffer$^+$ was added to each sample and vortexed at maximum speed until the pellet was resuspended. The suspension was incubated on ice for 30 min, during which was vortexed twice. The pellet was removed by centrifuging at top speed for 5 min and the supernatant was transferred to new tubes. 50μl of Luciferase Assay Buffer$^+$ was distributed into the wells of 96-well luciferase assay dish and 50μl of the cell extracts were added. The activity of the reporter gene (firefly luciferase) was measured on Luciferase Plate Reader by adding equal amounts ATP-supplemented Luciferin and measuring the bioluminescence in Relative Light Units (RLU).

5.2.4 DNA/Protein Interaction Assays

The Electrophoretic Mobility Shift Assay (EMSA) was used to analyze specific DNA binding activities in nuclear extracts of various cells *in vitro*.

5.2.4.1 Radioactive Labelling and Purification of DNA Probe

Oligonucleotides were dissolved in ddH$_2$O at 100 pmol/μl. For each probe 22,5 μl of sense and antisense oligonucleotides were mixed together with 5μl of 10xRed restriction enzyme buffer (MBI Fermentas) and incubated in a thermo block at 99°C for 1 min. Then the thermo block was switched off and left to slowly cool down till RT for hybridization of the oligos. Now the final volume obtained was 40 μl of 50 pmol/μl double stranded DNA. For labelling reaction hybridized duplexes were diluted till 20 ng/μl with 1xRed buffer (for dsDNA of 25 nucleotides in length, 1pmol corresponds to 9 ng, so 50 pmol/μl = 450 ng/μl and therefore dilution factor was 22,5x).

Typical composition of labelling reaction was:

ddH$_2$O	4,0 μl
10x PNK buffer	1,0 μl
ds-Oligo	2,0 μl (40 ng)
^{32}P-•ATP	2,0 μl (40 μCi)
PNK (T4 Polynucleotide kinase).........................	1,0 μl

The reaction mix was incubated for 30 min at 37°C. After addition of 5 µl of 6xAgarose buffer the samples were separated on 12% acrylamide/1xTBE PAAG against 1x TBE buffer for 1 hour. The gel was covered with plastic wrap, labeled with fluorescent markings and exposed to X-ray film for 1 min. Gel slices corresponding to the double-stranded oligonucleotides were cut and labeled. Probe was eluted with 150 µl of 1xRed buffer overnight while rotating.

Activity of labeled probe was estimated with a hand-held radioactivity monitor after spotting of 2 µl aliquot on a sheet of Whatman paper and diluted to approx. 20000 cpm/µl with 1xRed buffer. The probes were stored at –20°C and used within 3 weeks.

5.2.4.2 Electrophoretic Mobility Shift Assay (EMSA)

For EMSA, 6% PAA/0,4xTBE gels were prepared and allowed to polymerize for at least 1 hour at RT before pre-electrophoresis against 0,4x TBE at constant voltage 250 V till the current dropped from initial 30 mA to 10-11 mA.

DNA binding reactions were performed by mixing the ingredients listed below in the following order: water > buffer > poly dI/dC > Nuclear Extract > probe > Antibody

- 3x binding buffer – 5µl
- Poly dI/dC (1 µg/µl) – 1 µl
- Protein extract – 6-7µg (max 5µl) added.
- 1µl (1 µg/µl) of Antibody solution against the transcription factor to be studied added (optional).
- Radioactive probe (~20,000 cpm for each sample) – (1-2µl)
- H_2O till 15 µl

The mixture was incubated on ice for 15 min after the addition of the protein extract when radioactive probe was added, followed by 30 min incubation or addition of antibodies and further incubation for 1 hour on ice. Afterwards 10µl of binding reaction were loaded per slot. As a mobility marker 10µl 1x DNA gel loading buffer was loaded in one of the slots. The gel was ran again at 220 V for 3 – 3,5 hours, until the blue front reaches 5 cm from the end of the gel.

After separation of glass plates the gel was fixed in 10% acetic acid/10% Ethanol for 20 min, dried on Whatman 3MM paper in gel dryer and finally exposed for 1-2 days to X-ray film at -70°C using intensifying screen.

5.2.5 Mammalian Cell Cultures

5.2.5.1 Maintenance of Cell Lines

U2OS cells, Phoenix cells and NIH-3T3 cells and their derivatives were cultivated under sterile conditions in DMEM medium supplemented with all necessary ingredients (see above sub-chapter "Growth media") at 37°C and 5% CO_2. Cell cultures were split every day or every other day at appropriate densities to maintain the culture in logarithmically growing phase.

5.2.5.2 Splitting of Cell Cultures – Adherent Cells:

The growth media was aspirated and the cells washed once with pre-warmed sterile PBS. Then pre-warmed Trypsin-EDTA solution (2-3 ml per 10 cm culture plate) was added followed by incubation at 37°C for 5 min. The cells were detached from the bottom and resuspended by vigorously pipetting up and down with 1 ml micropipette until all cell aggregates disappear and uniformly dispersed cell suspension is formed. The trypsin was quenched by addition of 1-2 volumes of growth media and the cell number was determined by counting with haemocytometer. The required amount of cells was diluted in culture media, plated out and incubated at 37°C on strictly horizontal surface until the cells attach to the plastic (usually 0,5-1 h).

5.2.5.3 Cell Counting with Haemocytometer

To determine the number of cells adherent cultures were treated with trypsin as described above, appropriate dilutions in PBS/0,1% Trypan blue solution were made and dye-excluding living cells were counted under the microscope using Neubauer type haemocytometer chamber.

5.2.5.4 Crystal Violet Assay

NIH-3T3 cells or their derivatives were trypsinized, counted and after dilution plated at desired densities in multi-well tissue culture dishes, in the presence or absence of Doxycyline (2 µg/ml). Individual multi-well plate was dedicated for each sample, with at least 3 parallel samples per treatment, plated under identical conditions. Daily medium changes were performed during cultivation of the cells and one set of plates was removed daily for Crystal Violet staining.

Materials and Methods

The growth media was aspirated from the wells and the cells were stained with Crystal Violet solution. After 30 min of incubation at RT the plates were washed 10-times with tap water, air-dried for at least 30 min and stored until the end of the whole experiment. The dry plates were imaged for visual comparison using digital camera Konica-Minolta Dimage X1.

To determine the cell mass, Crystal Violet dye was extracted from the stained cell mass with 500 •l methanol for 30 min at RT with moderate shaking, 100µl transferred into a new 96-well flat bottom plate and optical density (OD) determined on spectrophotometer at • = 570 nm. The value of each well was designated as "relative cell mass" and used to construct cell growth curves.

5.2.5.5 Cell Transfections

The U2OS cells and Phoenix cells were transfected with the Calcium phosphate method. For NIH-3T3 cells and their derivatives both Calcium phosphate and dendrimer-based transfection methods were used.

5.2.5.5.1 Calcium Phosphate Transfection Method

The method described below is an optimisation of calcium phosphate transfection protocol, kindly provided by Dr. Carolin Schmittwolf.
Freshly split NIH-3T3 cells were used for the transfection. The evening before transfection 70-90% confluent NIH-3T3 culture was split between 1:10 and 1:15. Next morning growth media was changed 1-3 hrs before transfection. The reagents for different scales of cell culture transfection were pipetted according Table 5.2.

Table 5.2: DNA and reagent amounts for different surface areas

	DNA [µg]	Water (end vol.) [µl]	Ca-solution [µl]	2xHBS [µl]	Medium [ml]
10 cm plate	28	613	86,8	700	6
6-well dish	4,66	102	14,5	117	1
12-well dish	2,33	51	7,25	58,5	0,5

For each transfection were prepared DNA/Water/Ca-solution mix (solution A) and HBS in separate sterile tubes. Then solution A was slowly added (over 1 min) drop wise into the tube with HBS, under constant vortexing. The resulting mix was further incubated for 20 min and in drop wise fashion equally distributed over the cells. After 7-8 hours equal amount of fresh medium was added or if cytopathic effect was observed, the medium was completely changed.

5.2.5.5.2 Dendrimer-Based Transfection

The use of dendrimeric-based reagents PolyFect• and SuperFect• (Qiagen GmbH, Hilden, Germany) was optimized according to the manufacturer's protocols.

5.2.5.6 Selection of Transfected Cells

Selection of the cells was applied typically one day after transfection. The required selective agent was added to the culturing media in concentrations, varying for each cell type and each selective agent as shown on Table 5.3.

Table 5.3: Concentrations of selective agents for different cell types

	Puromycin	Neomycin
U2OS	8 µg/ml	N/A
293/Phoenix	1 µg/ml	550 µg/ml
NIH-3T3	5 µg/ml	0,6-1 mg/ml

The selective agent – supplemented media was changed each day to remove dead cells until the non-transfected control showed complete mortality. Then the selected cells were either used immediately (transient transfection) or selected further by cultivating in selective agent – supplemented media for up to three weeks (stable transfection).

5.2.5.7 Single Cell Cloning

The stable transfected cells were cultivated until multiple colonies were formed, then trypsinized and re-plated in the same culture dish. When sufficient density was achieved the cells were trypsinized again, diluted with media containing selective agent and plated out in flat-bottom 96-well dishes in densities 0,3 - 0,5 cells per well. The dishes were supplemented with fresh, selective agent-containing media regularly until isolated colonies originated from the single cells (1-3 weeks). Only dishes with frequency of colony appearance lower or equal than the calculated 0,3 – 0,5 cells per well were selected for further expansion in larger culture vessels.

5.2.5.8 Cryopreservation of Mammalian Cells

Exponentially growing cell culture was trypsinized as described above and the cells were settled by centrifugation. All further described procedures were done on ice. Each sample was resuspended in 1-1,5 ml freezing media (containing 90% FCS and 10% DMSO) and transferred to cryovial. The vials were placed in freezing box and left for freezing overnight at −70°C. The freezing box contains isopropanol which allows the cells to freeze slowly. The next day the vials were transferred to liquid nitrogen tank.

5.2.5.9 Thawing Mammalian Cells Out of Frozen Stock

The vials with the cells were brought from the liquid nitrogen storage on dry ice and thawed fast with continuous swirling in water bath at 37°C. The cell suspension was transferred to 50 ml tubes and 10 ml complete growth media was added slowly dropwise with constant mixing. Then the cells were settled by centrifuging at 1300 rpm, resuspended in fresh growth media and plated out.

5.2.6 Bacterial Manipulation

5.2.6.1 Cultivation of Bacteria (*E. coli*)

Bacterial cultures were maintained on solid or in liquid media. Single colonies were obtained by striking out on LB-agar plates. Miniprep DNA isolations were performed by using single colonies, inoculated in 2ml LB-ampicillin media and shake-cultured overnight at 37°C. For maxiprep DNA isolations 100µl from the mini-culture were inoculated in 250-500ml TB media and shake-cultured at 37°C overnight.

5.2.6.2 Preparation of E. coli Competent Bacterial Cells by $CaCl_2$ Method

A single bacterial colony was inoculated into 50ml of antibiotics-free LB broth and shake-cultured at 37°C overnight. Next day, 30 ml of the culture was diluted into 350 ml of antibiotics-free LB broth and the culturing continued under the same conditions. Every 30 minutes the OD595 of the culture was monitored until it reached 0,4 – 0,6. The culture was divided into six ice-cold 50 ml Falcon tubes and cooled on ice for 10 minutes. The bacteria were washed once in 20 ml of ice-cold 100 mM $CaCl_2$ solution. The washed bacteria were then incubated on ice for 30 minutes once in 50 ml of ice-cold $CaCl_2$ solution and once in 4 ml of ice-cold $CaCl_2$ solution. 0,2 ml aliquots of competent cell solution were frozen in 1,5 ml Eppendorf tubes and stored at −70°C.

Materials and Methods

5.2.6.3 Transformation of E.coli Competent Bacterial Cells

The competent bacterial cells were thawed on ice and divided in aliquots of 50 µl in sterile pre-cooled tubes. Estimated amounts of 1 ng and 10 ng of DNA were added in separate tubes, gently mixed and incubated on ice for 30 minutes. The competent cells were then heat-shocked at 42°C for exactly 90 seconds in water bath and cooled on ice for additional 2 minutes. 950 µl of sterile, antibiotics-free LB broth was added and the bacteria were shake-cultured at 37°C for one hour. Different volumes (50, 100, 200 µl) from the cultures were plated onto LB agar plates containing 50 µg/ml ampicillin and the plates were incubated at 37°C overnight.

5.2.6.4 Cryopreservation of Bacterial Cells

1-2 ml of overnight bacterial culture, grown in LB medium was centrifuged at 5000 rpm for 3 minutes at RT. The bacterial cells were collected as pellet, resuspended in 200-500 µl of cryoprotecting bacterial storage solution (80% medium and 20% glycerol) and stored in cryovials at −70°C.

5.2.7 Statistical Analyses

The experiments in this work are performed multiple times to ensure the reproducibility and reliability of the results. Any specifications about each particular experiment are denoted below the corresponding figure. Error bars represent the Standard Deviation. For GABP• •• overexpressing cells statistical significance is calculated as Homoscedastic T-test (two-sample equal variance) with two-tailed distribution. For Null hypothesis is taken the notion that the cell cultures with and without "treatment" grow with same speed. Any significant difference in the growth speeds proves the Null hypothesis wrong. Under "treatment" is meant addition of Doxycycline and respectively overexpression of GABP• ••. For the cells, constitutively overexpressing GABP• is used Paired T-test, also two-tailed distribution. In this case the same Null hypothesis is postulated, but here "treatment" means the presence of constitutively overexpressed GABP•. In both cases for calculating the statistical significance are taken the values of "treated" and "not treated" cell cultures from the time points, showing highest divergence from all of the experiments.

6. REFERENCES

Anderson KL, Smith KA, Conners K, McKercher SR, Maki RA, Torbett BE. (1998) Myeloid development is selectively disrupted in PU.1 null mice. *Blood.* 91: 3702-10.

Atlas E, Stramwasser M, Whiskin K, Mueller CR. (2000) GA-binding protein alpha/beta is a critical regulator of the BRCA1 promoter. *Oncogene.* 19: 1933-40.

Aurrekoetxea-Hernandez K, Buetti E. (2000) Synergistic action of GA-binding protein and glucocorticoid receptor in transcription from the mouse mammary tumor virus promoter. *J Virol* 74: 4988-98.

Avots A, Hoffmeyer A, Flory E, Cimanis A, Rapp UR, Serfling E. (1997) GABP factors bind to a distal interleukin 2 (IL-2) enhancer and contribute to c-Raf-mediated increase in IL-2 induction. *Mol Cell Biol* 17: 4381-9.

Bajaj G, Sharma RK. (2006) TNF-alpha-mediated cardiomyocyte apoptosis involves caspase-12 and calpain. *Biochem Biophys Res Commun* 345: 1558-64.

Bannert N, Avots A, Baier M, Serfling E, Kurth R. (1999) GA-binding protein factors, in concert with the coactivator CREB binding protein/p300, control the induction of the interleukin 16 promoter in T lymphocytes. *Proc Natl Acad Sci U S A* 96: 1541-6.

Bartel FO, Higuchi T, Spyropoulos DD. (2000) Mouse models in the study of the Ets family of transcription factors. *Oncogene* 19(55): 6443-54.

Bassuk AG and Leiden JM. (1997) The role of Ets transcription factors in the development and function of the mammalian immune system. *Adv Immunol* 64: 65-104.

Batchelor AH, Piper DE, de la Brousse FC, McKnight SL, Wolberger C. (1998) The structure of GABPalpha/beta: an ETS domain-ankyrin repeat heterodimer bound to DNA. *Science* 279:1037-41.

Bieche I, Tozlu S, Girault I, Onody P, Driouch K, Vidaud M, Lidereau R. (2004). Expression of PEA3/E1AF/ETV4, an Ets-related transcription factor, in breast tumors: positive links to MMP2, NRG1 and CGB expression. *Carcinogenesis* 25(3): 405-11.

Bister K, Nunn M, Moscovici C, Perbal B, Baluda M, Duesberg PH. (1982) Acute leukemia viruses E26 and avian myeloblastosis virus have related transformation-specific RNA sequences but different genetic structures, gene products, and oncogenic properties. *Proc Natl Acad Sci USA* 79: 3677-81.

Bitko V, Barik S. (2001) An endoplasmic reticulum-specific stress-activated caspase (caspase-12) is implicated in the apoptosis of A549 epithelial cells by respiratory syncytial virus. *J Cell Biochem* 80: 441-54.

Blobel GA. (2000) CREB-binding protein and p300: molecular integrators of hematopoietic transcription. *Blood* 95: 745-55.

Boccia LM, Lillicrap D, Newcombe K, Mueller CR. (1996) Binding of the Ets factor GA-binding protein to an upstream site in the factor IX promoter is a critical event in transactivation. *Mol Cell Biol* 16: 1929-35.

Bolwig GM, Bruder JT, Hearing P. (1992) Different binding site requirements for binding and activation for the bipartite enhancer factor EF-1A. *Nucleic Acids Res* 20: 6555-64.

Boone TC, Johnson MJ, De Clerck YA, Langley KE. (1990). cDNA cloning and expression of a metalloproteinase inhibitor related to tissue inhibitor of metalloproteinases. *Proc Natl Acad Sci U S A* 87(7): 2800-4.

Bottinger EP, Shelley CS, Farokhzad OC, Arnaout MA. (1994) The human beta 2 integrin CD18 promoter consists of two inverted Ets cis elements. *Mol Cell Biol* 14: 2604-15.

Bradford MM. (1976) A rapid and sensitive method for the quantitation of microgram quantities of protein utilizing the principle of protein-dye binding. *Anal Biochem* 72: 248-54.

Briguet A, Ruegg MA. (2000) The Ets transcription factor GABP is required for postsynaptic differentiation in vivo. *J Neurosci* 20: 5989-96.

Brown TA, McKnight SL. (1992) Specificities of protein-protein and protein-DNA interaction of GABP alpha and two newly defined ets-related proteins. *Genes Dev* 6: 2502-12.

Brummelkamp TR, Bernards R, Agami R. (2002) Stable suppression of tumorigenicity by virus-mediated RNA interference. *Cancer Cell* 2: 243-7.

Brummelkamp TR, Bernards R, Agami R. (2002) A system for stable expression of short interfering RNAs in mammalian cells. *Science* 296: 550-3.

References

Bush TS, St Coeur M, Resendes KK, Rosmarin AG. (2003) GA-binding protein (GABP) and Sp1 are required, along with retinoid receptors, to mediate retinoic acid responsiveness of CD18 (beta 2 leukocyte integrin): a novel mechanism of transcriptional regulation in myeloid cells. *Blood* 101: 311-7.

Campbell CE, Flenniken AM, Skup D, Williams BR (1991). Identification of a serum- and phorbol ester-responsive element in the murine tissue inhibitor of metalloproteinase gene. *J Biol Chem* 266(11): 7199-206.

Carter RS, Avadhani NG. (1994) Cooperative binding of GA-binding protein transcription factors to duplicated transcription initiation region repeats of the cytochrome c oxidase subunit IV gene. *J Biol Chem* 269: 4381-7.

Chinenov Y, Coombs C, Martin ME (2000) Isolation of a bi-directional promoter directing expression of the mouse GABPalpha and ATP synthase coupling factor 6 genes. *Gene* 261(2): 311-20.

Chinenov Y, Henzl M, Martin ME. (2000) The alpha and beta subunits of the GA-binding protein form a stable heterodimer in solution. Revised model of heterotetrameric complex assembly. *J Biol Chem* 275: 7749-56.

Chrast R, Chen H, Morris MA, Antonarakis SE. (1995) Mapping of the human transcription factor GABPA (E4TF1-60) gene to chromosome 21. *Genomics* 28: 119-22.

Cowley DO, Graves BJ. (2000) Phosphorylation represses Ets-1 DNA binding by reinforcing autoinhibition. *Genes Dev* 14: 366-76.

Crook MF, Olive M, Xue HH, Langenickel TH, Boehm M, Leonard WJ, Nabel EG. (2007) GA-binding protein regulates KIS gene expression, cell migration, and cell cycle progression. *FASEB J*.

Curcić D, Glibetić M, Larson DE, Sells BH. (1997) GA-binding protein is involved in altered expression of ribosomal protein L32 gene. *J Cell Biochem.* 65: 287-307.

Day CJ, Kim MS, Stephens SR, Simcock WE, Aitken CJ, Nicholson GC, Morrison NA. (2004) Gene array identification of osteoclast genes: differential inhibition of osteoclastogenesis by cyclosporin A and granulocyte macrophage colony stimulating factor. *J Cell Biochem* 91: 303-15.

de la Brousse FC, Birkenmeier EH, King DS, Rowe LB, McKnight SL. (1994) Molecular and genetic characterization of GABP beta. *Genes Dev* 8: 1853-65.

Diaz-Horta O, Kamagate A, Herchuelz A, Van Eylen F. (2002) Na/Ca exchanger overexpression induces endoplasmic reticulum-related apoptosis and caspase-12 activation in insulin-releasing BRIN-BD11 cells. *Diabetes* 51: 1815-24.

Dittmer J, Nordheim A. (1998) Ets transcription factors and human disease. *Biochim Biophys Acta* 1377: F1-11.

Donaldson LW, Petersen JM, Graves BJ, McIntosh LP. (1996) Solution structure of the ETS domain from murine Ets-1: a winged helix-turn-helix DNA binding motif. *EMBO J* 15: 125-34.

Douville P, Hagmann M, Georgiev O, Schaffner W. (1995) Positive and negative regulation at the herpes simplex virus ICP4 and ICP0 TAATGARAT motifs. *Virology.* 207: 107-16.

Du K, Leu JI, Peng Y, Taub R. (1998) Transcriptional up-regulation of the delayed early gene HRS/SRp40 during liver regeneration. Interactions among YY1, GA-binding proteins, and mitogenic signals. *J Biol Chem* 273: 35208-15.

Duclert A, Savatier N, Schaeffer L, Changeux JP. (1996) Identification of an element crucial for the sub-synaptic expression of the acetylcholine receptor epsilon-subunit gene. *J Biol Chem* 271: 17433-8.

Fitzsimmons D, Hodsdon W, Wheat W, Maira SM, Wasylyk B, Hagman J. (1996). Pax-5 (BSAP) recruits Ets proto-oncogene family proteins to form functional ternary complexes on a B-cell-specific promoter. *Genes Dev* 10(17): 2198-211.

Flory E, Hoffmeyer A, Smola U, Rapp UR, Bruder JT. (1996) Raf-1 kinase targets GA-binding protein in transcriptional regulation of the human immunodeficiency virus type 1 promoter. *J Virol.* 70: 2260-8.

Fromm L, Burden SJ. (1998) Synapse-specific and neuregulin-induced transcription require an ets site that binds GABPalpha/GABPbeta. *Genes Dev.* 12: 3074-83.

Fromm L, Burden SJ. (2001) Neuregulin-1-stimulated phosphorylation of GABP in skeletal muscle cells. *Biochemistry* 40: 5306-12.

Fromm L, Rhode M. (2004) Neuregulin-1 induces expression of Egr-1 and activates acetylcholine receptor transcription through an Egr-1-binding site. *J Mol Biol* 339: 483-94.

Fujita E, Kouroku Y, Jimbo A, Isoai A, Maruyama K, Momoi T. (2002) Caspase-12 processing and fragment translocation into nuclei of tunicamycin-treated cells. *Cell Death Differ* 9: 1108-14.

References

Fujita J, Crane AM, Souza MK, Dejosez M, Kyba M, Flavell RA, Thomson JA, Zwaka TP. (2008) Caspase activity mediates the differentiation of embryonic stem cells. *Cell Stem Cell* Vol 2:595-601.

Gaines P, Berliner N. (2003) Retinoids in myelopoiesis. *J Biol Regul Homeost Agents* 17: 46-65.

Gajewski KM, Schulz RA. (1995) Requirement of the ETS domain transcription factor D-ELG for egg chamber patterning and development during Drosophila oogenesis. *Oncogene* 11: 1033-40.

Galang CK, Muller WJ, Foos G, Oshima RG, Hauser CA. (2004). Changes in the expression of many Ets family transcription factors and of potential target genes in normal mammary tissue and tumors. *J Biol Chem* 279(12): 11281-92.

Galvagni F, Capo S, Oliviero S. (2001) Sp1 and Sp3 physically interact and co-operate with GABP for the activation of the utrophin promoter. *J Mol Biol* 306: 985-96.

Garvie CW, Hagman J, Wolberger C. (2001). Structural studies of Ets-1/Pax5 complex formation on DNA. *Mol Cell* 8(6): 1267-76.

Garvie CW, Pufall MA, Graves BJ, Wolberger C. (2002) Structural analysis of the autoinhibition of Ets-1 and its role in protein partnerships. *J Biol Chem.* Nov 22; 277(47):45529-36.

Geng Y. and Johnson L.F. (1993). Lack of an initiator element is responsible for multiple transcriptional initiation sites of the TATA-less mouse thymidylate synthase promoter. Mol Cell Biol 13(8): 4894-903.

Genuario RR, Kelley DE, Perry RP. (1993) Comparative utilization of transcription factor GABP by the promoters of ribosomal protein genes rpL30 and rpL32. *Gene Expr* 3: 279-88.

Genuario RR, Perry RP. (1996) The GA-binding protein can serve as both an activator and repressor of ribosomal protein gene transcription. *J Biol Chem* 271: 4388-95.

Ghosh A, Kolodkin AL. (1998) Specification of neuronal connectivity:ETS marks the spot. *Cell* 95:303-6.

Goto M, Shimizu T, Sawada J, Sawa C, Watanabe H, Ichikawa H, Ohira M, Ohki M, Handa H. (1995) Assignment of the E4TF1-60 gene to human chromosome 21q21.2-q21.3. *Gene* 166: 337-8.

Gramolini AO, Angus LM, Schaeffer L, Burton EA, Tinsley JM, Davies KE, Changeux JP, Jasmin BJ. (1999) Induction of utrophin gene expression by heregulin in skeletal muscle cells: role of the N-box motif and GA binding protein. *Proc Natl Acad Sci U S A* 96: 3223-7.

Graves BJ, Petersen JM. (1998) Specificity within the ets family of transcription factors. *Adv Cancer Res.* 75: 1-55.

Graves BJ. (1998) Inner workings of a transcription factor partnership. *Science* 279: 1000-2.

Gugneja S, Virbasius CM, Scarpulla RC. (1996) Nuclear respiratory factors 1 and 2 utilize similar glutamine-containing clusters of hydrophobic residues to activate transcription. *Mol Cell Biol* 16: 5708-16.

Gugneja S, Virbasius JV, Scarpulla RC. (1995) Four structurally distinct, non-DNA-binding subunits of human nuclear respiratory factor 2 share a conserved transcriptional activation domain. *Mol Cell Biol* 15: 102-11.

Gutman A. and B. Wasylyk (1990). The collagenase gene promoter contains a TPA and oncogene-responsive unit encompassing the PEA3 and AP-1 binding sites. *EMBO J* 9(7): 2241-6.

Gyrd-Hansen M, Krag TO, Rosmarin AG, Khurana TS. (2002) Sp1 and the ets-related transcription factor complex GABP alpha/beta functionally cooperate to activate the utrophin promoter. *J Neurol Sci* 197: 27-35.

Hagman J, Grosschedl R. (1992) An inhibitory carboxyl-terminal domain in Ets-1 and Ets-2 mediates differential binding of ETS family factors to promoter sequences of the mb-1 gene. *Proc Natl Acad Sci U S A* 89: 8889-93.

Handschin C, Kobayashi YM, Chin S, Seale P, Campbell KP, Spiegelman BM. (2007) PGC-1alpha regulates the neuromuscular junction program and ameliorates Duchenne muscular dystrophy. *Genes Dev* 21: 770-83.

Hao H, Qi H, Ratnam M. (2003) Modulation of the folate receptor type beta gene by coordinate actions of retinoic acid receptors at activator Sp1/ets and repressor AP-1 sites. *Blood.* 101: 4551-60.

Hart AH, Reventar R, Bernstein A. (2000) Genetic analysis of ETS genes in C. elegans. *Oncogene* 19: 6400-8.

Hasan NM, MacDonald MJ. (2002) Sp/Kruppel-like transcription factors are essential for the expression of mitochondrial glycerol phosphate dehydrogenase promoter B. *Gene* 296: 221-34.

Hauck L, Kaba RG, Lipp M, Dietz R, von Harsdorf R. (2002) Regulation of E2F1-dependent gene transcription and apoptosis by the ETS-related transcription factor GABPgamma1. *Mol Cell Biol* 22: 2147-58.

References

Hempel N, Wang H, LeCluyse EL, McManus ME, Negishi M. (2004) The human sulfotransferase SULT1A1 gene is regulated in a synergistic manner by Sp1 and GA binding protein. *Mol Pharmacol* 66: 1690-701.

Hickstein DD, Hickey MJ, Collins SJ. (1988) Transcriptional regulation of the leukocyte adherence protein beta subunit during human myeloid cell differentiation. *J Biol Chem* 263: 13863-7.

Hoare S, Copland JA, Wood TG, Jeng YJ, Izban MG, Soloff MS. (1999) Identification of a GABP alpha/beta binding site involved in the induction of oxytocin receptor gene expression in human breast cells, potentiation by c-Fos/c-Jun. *Endocrinology* 140: 2268-79.

Hoffmeyer A, Avots A, Flory E, Weber CK, Serfling E, Rapp UR. (1998) The GABP-responsive element of the interleukin-2 enhancer is regulated by JNK/SAPK-activating pathways in T lymphocytes. *J Biol Chem* 273: 10112-9.

Huhtala P, Chow LT, Tryggvason K, (1990). Structure of the human type IV collagenase gene. *J Biol Chem* 265(19): 11077-82.

Ikeda S, Mochizuki A, Sarker AH, Seki S. (2000). Identification of functional elements in the bidirectional promoter of the mouse Nthl1 and Tsc2 genes. *Biochem Biophys Res Commun* 273(3): 1063-8.

Imaki H, Nakayama K, Delehouzee S, Handa H, Kitagawa M, Kamura T, Nakayama KI. (2003) Cell cycle-dependent regulation of the Skp2 promoter by GA-binding protein. *Cancer Res* 63: 4607-13.

Izumi M, Yokoi M, Nishikawa NS, Miyazawa H, Sugino A, Yamagishi M, Yamaguchi M, Matsukage A, Yatagai F, Hanaoka F. (2000) Transcription of the catalytic 180-kDa subunit gene of mouse DNA polymerase alpha is controlled by E2F, an Ets-related transcription factor, and Sp1. *Biochim Biophys Acta* 1492: 341-52.

Janzen V, Fleming HE, Riedt T, Karlsson G, Riese MJ, Lo Celso C, Reynolds G, Milne CD, Paige CJ, Karlsson S, Woo M, Scadden DT. Hematopoietic stem cell responsiveness to exogenous signals is limited by caspase-3. *Cell Stem Cell* Vol. 2: 584-94.

Jaworski A, Smith CL, Burden SJ. (2007) GA-binding protein is dispensable for neuromuscular synapse formation and synapse-specific gene expression. *Mol Cell Biol* 27: 5040-6.

Jeong BC, Kim MY, Lee JH, Kee HJ, Kho DH, Han KE, Qian YR, Kim JK, Kim KK. (2006) Brain-specific angiogenesis inhibitor 2 regulates VEGF through GABP that acts as a transcriptional repressor. *FEBS Lett* 580: 669-76.

Jiang P, Kumar A, Parrillo JE, Dempsey LA, Platt JL, Prinz RA, Xu X. (2002) Cloning and characterization of the human heparanase-1 (HPR1) gene promoter: role of GA-binding protein and Sp1 in regulating HPR1 basal promoter activity. *J Biol Chem* 277: 8989-98.

Jonsen MD, Petersen JM, Xu QP, Graves BJ. (1996) Characterization of the cooperative function of inhibitory sequences in Ets-1. *Mol Cell Biol* 16: 2065-73.

Jordan R, Wang L, Graczyk TM, Block TM, Romano PR. (2002) Replication of a cytopathic strain of bovine viral diarrhea virus activates PERK and induces endoplasmic reticulum stress-mediated apoptosis of MDBK cells. J Virol 76: 9588-99.

Jousset C, Carron C, Boureux A, Quang CT, Oury C, Dusanter-Fourt I, Charon M, Levin J, Bernard O, Ghysdael J. (1997) A domain of TEL conserved in a subset of ETS proteins defines a specific oligomerization interface essential to the mitogenic properties of the TEL-PDGFR beta oncoprotein. EMBO J 16: 69-82.

Kalai M, Lamkanfi M, Denecker G, Boogmans M, Lippens S, Meeus A, Declercq W, Vandenabeele P. (2003) Regulation of the expression and processing of caspase-12. J Cell Biol 162: 457-67.

Kamura T, Handa H, Hamasaki N, Kitajima S. (1997) Characterization of the human thrombopoietin gene promoter. A possible role of an Ets transcription factor, E4TF1/GABP. *J Biol Chem* 272: 11361-8.

Khurana TS, Davies KE. (2003) Pharmacological strategies for muscular dystrophy. *Nat Rev Drug Discov* 2: 379-90.

Khurana TS, Rosmarin AG, Shang J, Krag TO, Das S, Gammeltoft S. (1999) Activation of utrophin promoter by heregulin via the ets-related transcription factor complex GA-binding protein alpha/beta. Mol Biol Cell 10: 2075-86.

Kilic M, Schafer R, Hoppe J, Kagerhuber U. (2002) Formation of noncanonical high molecular weight caspase-3 and -6 complexes and activation of caspase-12 during serum starvation induced apoptosis in AKR-2B mouse fibroblasts. Cell Death Differ 9: 125-37.

Kinoshita K, Ura H, Akagi T, Usuda M, Koide H, Yokota T. (2007) GABPalpha regulates Oct-3/4 expression in mouse embryonic stem cells. *Biochem Biophys Res Commun* 353: 686-91.

References

Koike S, Schaeffer L, Changeux JP. (1995) Identification of a DNA element determining synaptic expression of the mouse acetylcholine receptor delta-subunit gene. *Proc Natl Acad Sci U S A* 92: 10624-8.

Kung AL, Rebel VI, Bronson RT, Ch'ng LE, Sieff CA, Livingston DM, Yao TP. (2000) Gene dose-dependent control of hematopoiesis and hematologic tumor suppression by CBP. *Genes Dev* 14: 272-7.

Laemmli UK. (1970) Cleavage of structural proteins during the assembly of the head of bacteriophage T4. *Nature* 227: 680-5.

LaMarco K, Thompson CC, Byers BP, Walton EM, McKnight SL. (1991) Identification of Ets- and notch-related subunits in GA binding protein. *Science* 253: 789-92.

Laudet V, Hanni C, Stehelin D, Duterque-Coquillaud M. (1999) Molecular phylogeny of the ETS gene family. *Oncogene* 18: 1351-9.

Li XR, Chong AS, Wu J, Roebuck KA, Kumar A, Parrillo JE, Rapp UR, Kimberly RP, Williams JW, Xu X. (1999) Transcriptional regulation of Fas gene expression by GA-binding protein and AP-1 in T cell antigen receptor.CD3 complex-stimulated T cells. *J Biol Chem* 274: 35203-10.

Lim F, Kraut N, Framptom J, Graf T. (1992) DNA binding by c-Ets-1, but not v-Ets, is repressed by an intramolecular mechanism. *EMBO J* 11: 643-52.

Lucas ME, Crider KS, Powell DR, Kapoor-Vazirani P, Vertino PM. (2009). Methylation-sensitive regulation of TMS1/ASC by the ETS factor, GA-binding protein-alpha. *J Biol Chem*. Mar 25. [Epub ahead of print]

Luo M, Shang J, Yang Z, Simkevich CP, Jackson CL, King TC, Rosmarin AG. (1999) Characterization and localization to chromosome 7 of psihGABPalpha, a human processed pseudogene related to the ets transcription factor, hGABPalpha. *Gene* 234: 119-26.

Macdonald G, Stramwasser M, Mueller CR. (2007) Characterization of a negative transcriptional element in the BRCA1 promoter. *Breast Cancer Res* 9: R49.

Marchioni M, Morabito S, Salvati AL, Beccari E, Carnevali F. (1993) XrpFI, an amphibian transcription factor composed of multiple polypeptides immunologically related to the GA-binding protein alpha and beta subunits, is differentially expressed during Xenopus laevis development. *Mol Cell Biol* 13: 6479-89.

Markiewicz S, Bosselut R, Le Deist F, de Villartay JP, Hivroz C, Ghysdael J, Fischer A, de Saint Basile G. (1996) Tissue-specific activity of the gammac chain gene promoter depends upon an Ets binding site and is regulated by GA-binding protein. *J Biol Chem*. 271: 14849-55.

Martin ME, Chinenov Y, Yu M, Schmidt TK, Yang XY. (1996) Redox regulation of GA-binding protein-alpha DNA binding activity. *J Biol Chem* 271: 25617-23.

McLean TW, Ringold S, Neuberg D, Stegmaier K, Tantravahi R, Ritz J, Koeffler HP, Takeuchi S, Janssen JW, Seriu T, Bartram CR, Sallan SE, Gilliland DG, Golub TR. (1996). TEL/AML-1 dimerizes and is associated with a favorable outcome in childhood acute lymphoblastic leukemia. *Blood* 88(11): 4252-8.

Mootha VK, Handschin C, Arlow D, Xie X, St Pierre J, Sihag S, Yang W, Altshuler D, Puigserver P, Patterson N, Willy PJ, Schulman IG, Heyman RA, Lander ES, Spiegelman BM. (2004) Erralpha and Gabpa/b specify PGC-1alpha-dependent oxidative phosphorylation gene expression that is altered in diabetic muscle. *Proc Natl Acad Sci U S A* 101: 6570-5.

Morii E, Ogihara H, Oboki K, Sawa C, Sakuma T, Nomura S, Esko JD, Handa H, Kitamura Y. (2001) Inhibitory effect of the mi transcription factor encoded by the mutant mi allele on GA binding protein-mediated transcript expression in mouse mast cells. *Blood* 97: 3032-9.

Morishima N, Nakanishi K, Takenouchi H, Shibata T, Yasuhiko Y. (2002) An endoplasmic reticulum stress-specific caspase cascade in apoptosis. Cytochrome c-independent activation of caspase-9 by caspase-12. *J Biol Chem* 277: 34287-94.

Mueller BU, Pabst T, Osato M, Asou N, Johansen LM, Minden MD, Behre G, Hiddemann W, Ito Y, Tenen DG. (2002) Heterozygous PU.1 mutations are associated with acute myeloid leukemia. *Blood* 100: 998-1007.

Mundle SD, Saberwal G. (2003) Evolving intricacies and implications of E2F1 regulation. *FASEB J* 17: 569-74.

Nakagawa T, Zhu H, Morishima N, Li E, Xu J, Yankner BA, Yuan J. (2000) Caspase-12 mediates endoplasmic-reticulum-specific apoptosis and cytotoxicity by amyloid-beta. *Nature* 403: 98-103.

Neekhra A, Luthra S, Chwa M, Seigel G, Gramajo AL, Kuppermann BD, Kenney MC. (2007) Caspase-8, -12, and -3 activation by 7-ketocholesterol in retinal neurosensory cells. *Invest Ophthalmol Vis Sci* 48: 1362-7.

References

Nickel J, Short ML, Schmitz A, Eggert M, Renkawitz R. (1995) Methylation of the mouse M-lysozyme downstream enhancer inhibits heterotetrameric GABP binding. Nucleic Acids Res 23: 4785-92.

Nishikawa N, Izumi M, Yokoi M, Miyazawa H, Hanaoka F. (2001) E2F regulates growth-dependent transcription of genes encoding both catalytic and regulatory subunits of mouse primase. Genes Cells. 6: 57-70.

Nuchprayoon I, Simkevich CP, Luo M, Friedman AD, Rosmarin AG. (1997) GABP cooperates with c-Myb and C/EBP to activate the neutrophil elastase promoter. Blood 89: 4546-54.

Nunn M, Weiher H, Bullock P, Duesberg P. (1984) Avian erythroblastosis virus E26: nucleotide sequence of the tripartite onc gene and of the LTR, and analysis of the cellular prototype of the viral ets sequence. Virology 139: 330-9.

Nunn MF, Seeburg PH, Moscovici C, Duesberg PH. (1983) Tripartite structure of the avian erythroblastosis virus E26 transforming gene. Nature 306: 391-5.

Obika S, Reddy SY, Bruice TC. (2003) Sequence specific DNA binding of Ets-1 transcription factor: molecular dynamics study on the Ets domain-DNA complexes. J Mol Biol. Aug 8; 331(2):345-59.

Oelgeschlager M, Nuchprayoon I, Luscher B, Friedman AD. (1996) C/EBP, c-Myb, and PU.1 cooperate to regulate the neutrophil elastase promoter. Mol Cell Biol 16: 4717-25.

Ogawa K, Burbelo PD, Sasaki M, Yamada Y. (1988). The laminin B2 chain promoter contains unique repeat sequences and is active in transient transfection. J Biol Chem 263(17): 8384-9.

Oikawa T, Yamada T. (2003) Molecular biology of the Ets family of transcription factors. Gene 303: 11-34.

Okada Y, Yano K, Jin E, Funahashi N, Kitayama M, Doi T, Spokes K, Beeler DL, Shih SC, Okada H, Danilov TA, Maynard E, Minami T, Oettgen P, Aird WC. (2007) A three-kilobase fragment of the human Robo4 promoter directs cell type-specific expression in endothelium. Circ Res 100: 1712-22.

O'Leary DA, Koleski D, Kola I, Hertzog PJ, Ristevski S. (2005) Identification and expression analysis of alternative transcripts of the mouse GA-binding protein (Gabp) subunits alpha and beta1. Gene 344: 79-92.

O'Leary DA, Noakes PG, Lavidis NA, Kola I, Hertzog PJ, Ristevski S. (2007) Targeting of the ETS factor GABPalpha disrupts neuromuscular junction synaptic function. Mol Cell Biol 27: 3470-80.

O'Leary DA, Pritchard MA, Xu D, Kola I, Hertzog PJ, Ristevski S. (2004) Tissue-specific overexpression of the HSA21 gene GABPalpha: implications for DS. Biochim Biophys Acta 1739: 81-7.

Ouyang L, Jacob KK, Stanley FM (1996). GABP mediates insulin-increased prolactin gene transcription. J Biol Chem. 271: 10425-8.

Papas TS; Watson DK; Sacchi N; Fujiwara S; Seth AK; Fisher RJ; Bhat NK; Mavrothalassitis G; Koizumi S; Jorcyk CL (1990). ETS family of genes in leukemia and Down syndrome. Am J Med Genet Suppl 7: 251-61.

Patton J, Block S, Coombs C, Martin ME (2006). Identification of functional elements in the murine Gabp alpha/ATP synthase coupling factor 6 bi-directional promoter. Gene 369: 35-44.

Perry R. P. (2005). The architecture of mammalian ribosomal protein promoters. BMC Evol Biol 5(1): 15.

Petersen JM, Skalicky JJ, Donaldson LW, McIntosh LP, Alber T, Graves BJ. (1995) Modulation of transcription factor Ets-1 DNA binding: DNA-induced unfolding of an alpha helix. Science 269: 1866-9.

Pierson III LS and Kennedy C. The Mechanisms of Gene Regulation. Microbial Genetics Lecture Notes, developed for a class at the University of Arizona.

Pierson LSlaCK. The Mechanisms of Gene Regulation - Microbial Genetics Lecture Notes.

Puel A, Ziegler SF, Buckley RH, Leonard WJ. (1998) Defective IL7R expression in T(-) B(+)NK(+) severe combined immunodeficiency. Nat Genet 20: 394-7.

QIAGEN. (2000) PolyFect® Transfection Reagent Handbook.

QIAGEN. (2002) SuperFect® Transfection Reagent Handbook.

Rahman A, Esmaili A, Saatcioglu F. (1995) A unique thyroid hormone response element in the human immunodeficiency virus type 1 long terminal repeat that overlaps the Sp1 binding sites. J Biol Chem. 270: 31059-64.

Rao MK, Maiti S, Ananthaswamy HN, Wilkinson MF. (2002) A highly active homeobox gene promoter regulated by Ets and Sp1 family members in normal granulosa cells and diverse tumor cell types. J Biol Chem 277: 26036-45.

References

Rao RV, Castro-Obregon S, Frankowski H, Schuler M, Stoka V, del Rio G, Bredesen DE, Ellerby HM. (2002) Coupling endoplasmic reticulum stress to the cell death program. An Apaf-1-independent intrinsic pathway. J Biol Chem 277: 21836-42.

Rao RV, Hermel E, Castro-Obregon S, del Rio G, Ellerby LM, Ellerby HM, Bredesen DE. (2001) Coupling endoplasmic reticulum stress to the cell death program. Mechanism of caspase activation. J Biol Chem 276: 33869-74.

Reeves RH, Cabin DE. (1999) Mouse chromosome 16. Mamm Genome 10: 957.

Reisner AH, Nemes P, Bucholtz C. (1975) The use of Coomassie Brilliant Blue G250 perchloric acid solution for staining in electrophoresis and isoelectric focusing on polyacrylamide gels. Anal Biochem 64: 509-16.

Ristevski S, O'Leary DA, Thornell AP, Owen MJ, Kola I, Hertzog PJ. (2004) The ETS transcription factor GABPalpha is essential for early embryogenesis. *Mol Cell Biol* 24: 5844-9.

Rosen GD, Barks JL, Iademarco MF, Fisher RJ, Dean DC. (1994) An intricate arrangement of binding sites for the Ets family of transcription factors regulates activity of the alpha 4 integrin gene promoter. *J Biol Chem* 269: 15652-60.

Rosmarin AG, Karen K. Resendes, Zhongfa Yang, John N. McMillan, and Shawna L. Fleming. (2004) GA-binding protein transcription factor: a review of GABP as an integrator of intracellular signaling and protein–protein interactions. *Blood Cells, Molecules, and Diseases*: 143-54.

Rosmarin AG, Caprio D, Levy R, Simkevich C. (1995) CD18 (beta 2 leukocyte integrin) promoter requires PU.1 transcription factor for myeloid activity. *Proc Natl Acad Sci U S A* 92: 801-5.

Rosmarin AG, Caprio DG, Kirsch DG, Handa H, Simkevich CP. (1995) GABP and PU.1 compete for binding, yet cooperate to increase CD18 (beta 2 leukocyte integrin) transcription. *J Biol Chem* 270: 23627-33.

Rosmarin AG, Luo M, Caprio DG, Shang J, Simkevich CP. (1998) Sp1 cooperates with the ets transcription factor, GABP, to activate the CD18 (beta2 leukocyte integrin) promoter. *J Biol Chem* 273: 13097-103.

Rosmarin AG, Weil SC, Rosner GL, Griffin JD, Arnaout MA, Tenen DG. (1989) Differential expression of CD11b/CD18 (Mo1) and myeloperoxidase genes during myeloid differentiation. *Blood* 73: 131-6.

Rudge TL, Johnson LF. (2002) Synergistic activation of the TATA-less mouse thymidylate synthase promoter by the Ets transcription factor GABP and Sp1. Exp Cell Res 274: 45-55.

Sadasivan E, Cedeno MM, Rothenberg SP. (1994) Characterization of the gene encoding a folate-binding protein expressed in human placenta. Identification of promoter activity in a G-rich SP1 site linked with the tandemly repeated GGAAG motif for the ets encoded GA-binding protein. J Biol Chem 269: 4725-35.

Sanges D, Marigo V. (2006) Cross-talk between two apoptotic pathways activated by endoplasmic reticulum stress: differential contribution of caspase-12 and AIF. Apoptosis 11: 1629-41.

Savoysky E, Mizuno T, Sowa Y, Watanabe H, Sawada J, Nomura H, Ohsugi Y, Handa H, Sakai T. (1994) The retinoblastoma binding factor 1 (RBF-1) site in RB gene promoter binds preferentially E4TF1, a member of the Ets transcription factors family. Oncogene 9: 1839-46.

Sawa C, Goto M, Suzuki F, Watanabe H, Sawada J, Handa H. (1996) Functional domains of transcription factor hGABP beta1/E4TF1-53 required for nuclear localization and transcription activation. *Nucleic Acids Res* 24: 4954-61.

Sawa C, Yoshikawa T, Matsuda-Suzuki F, Delehouzee S, Goto M, Watanabe H, Sawada J, Kataoka K, Handa H. (2002) YEAF1/RYBP and YAF-2 are functionally distinct members of a cofactor family for the YY1 and E4TF1/hGABP transcription factors. J Biol Chem 277: 22484-90.

Sawada J, Goto M, Sawa C, Watanabe H, Handa H. (1994) Transcriptional activation through the tetrameric complex formation of E4TF1 subunits. *EMBO J* 13: 1396-402.

Sawada J, Goto M, Sawa C, Watanabe H, Handa H. (1994) Transcriptional activation through the tetrameric complex formation of E4TF1 subunits. *EMBO J* 13: 1396-402.

Sawada J, Simizu N, Suzuki F, Sawa C, Goto M, Hasegawa M, Imai T, Watanabe H, Handa H. (1999) Synergistic transcriptional activation by hGABP and select members of the activation transcription factor/cAMP response element-binding protein family. J Biol Chem 274: 35475-82.

Scarpulla RC. (2002) Nuclear activators and coactivators in mammalian mitochondrial biogenesis. *Biochim Biophys Acta* 1576: 1-14.

Schaeffer L, Duclert N, Huchet-Dymanus M, Changeux JP. (1998) Implication of a multisubunit Ets-related transcription factor in synaptic expression of the nicotinic acetylcholine receptor. *EMBO J* 17: 3078-90.

References

Schweppe R, Gutierrez-Hartmann A. (2001) Pituitary Ets-1 and GABP bind to the growth factor regulatory sites of the rat prolactin promoter. *Nucleic Acids Res.* 29: 1251-60.

Schwerk C, Schulze-Osthoff K. (2003) Non-apoptotic functions of caspases in cellular proliferation and differentiation. *Biochem Pharmacol* 66: 1453-8.

Scott EW, Simon MC, Anastasi J, Singh H. (1994) Requirement of transcription factor PU.1 in the development of multiple hematopoietic lineages. *Science* 265: 1573-7.

Sedgwick SG, Smerdon SJ. (1999) The ankyrin repeat: a diversity of interactions on a common structural framework. *Trends Biochem Sci* 24: 311-6.

Seelan RS, Gopalakrishnan L, Scarpulla RC, Grossman LI. (1996) Cytochrome c oxidase subunit VIIa liver isoform. Characterization and identification of promoter elements in the bovine gene. *J Biol Chem* 271: 2112-20.

Seelan RS, Grossman LI. (1997) Structural organization and promoter analysis of the bovine cytochrome c oxidase subunit VIIc gene. A functional role for YY1. *J Biol Chem* 272: 10175-81.

Sharrocks AD. (2001) The ETS-domain transcription factor family. *Nat Rev Mol Cell Biol* 2: 827-37.

Sharrocks AD, Brown AL, Ling Y, Yates PR. (1997) The ETS-domain transcription factor family. *Int J Biochem Cell Biol* 29: 1371-87.

Shiio Y, Sawada J, Handa H, Yamamoto T, Inoue J. (1996) Activation of the retinoblastoma gene expression by Bcl-3: implication for muscle cell differentiation. *Oncogene* 12: 1837-45.

Shiraishi H, Okamoto H, Yoshimura A, Yoshida H. (2006) ER stress-induced apoptosis and caspase-12 activation occurs downstream of mitochondrial apoptosis involving Apaf-1. *J Cell Sci* 119: 3958-66.

Shore P and Sharrocks AD. (1995). The ETS-domain transcription factors Elk-1 and SAP-1 exhibit differential DNA binding specificities. *Nucleic Acids Res* 23(22): 4698-706.

Skalicky JJ, Donaldson LW, Petersen JM, Graves BJ, McIntosh LP. (1996) Structural coupling of the inhibitory regions flanking the ETS domain of murine Ets-1. *Protein Sci* 5: 296-309.

Sowa Y, Shiio Y, Fujita T, Matsumoto T, Okuyama Y, Kato D, Inoue J, Sawada J, Goto M, Watanabe H, Handa H, Sakai T. (1997) Retinoblastoma binding factor 1 site in the core promoter region of the human RB gene is activated by hGABP/E4TF1. *Cancer Res.* 57: 3145-8.

Sucharov C, Basu A, Carter RS, Avadhani NG. (1995) A novel transcriptional initiator activity of the GABP factor binding ets sequence repeat from the murine cytochrome c oxidase Vb gene. *Gene Expr* 5: 93-111.

Sun W, Graves BJ, Speck NA. (1995) Transactivation of the Moloney murine leukemia virus and T-cell receptor beta-chain enhancers by cbf and ets requires intact binding sites for both proteins. *J Virol* 69: 4941-9.

Sunesen M, Huchet-Dymanus M, Christensen MO, Changeux JP. (2003) Phosphorylation-elicited quaternary changes of GA binding protein in transcriptional activation. *Mol Cell Biol* 23: 8008-18.

Takahashi Y, Kako K, Arai H, Ohishi T, Inada Y, Takehara A, Fukamizu A, Munekata E. (2002) Characterization and identification of promoter elements in the mouse COX17 gene. *Biochim Biophys Acta* 1574: 359-64.

Tanaka M, Ueda A, Kanamori H, Ideguchi H, Yang J, Kitajima S, Ishigatsubo Y. (2002) Cell-cycle-dependent regulation of human aurora A transcription is mediated by periodic repression of E4TF1. *J Biol Chem* 277: 10719-26.

Thompson CC, Brown TA, McKnight SL. (1991) Convergence of Ets- and notch-related structural motifs in a heteromeric DNA binding complex. *Science* 253: 762-8.

Tomaras GD, Foster DA, Burrer CM, Taffet SM. (1999) ETS transcription factors regulate an enhancer activity in the third intron of TNF-alpha. *J Leukoc Biol* 66: 183-93.

Trojanowska, M. (2000). Ets factors and regulation of the extracellular matrix. *Oncogene* 19(55):6464-71.

Vassias I, Hazan U, Michel Y, Sawa C, Handa H, Gouya L, Morinet F. (1998) Regulation of human B19 parvovirus promoter expression by hGABP (E4TF1) transcription factor. *J Biol Chem* 273: 8287-93.

Verhoef K, Sanders RW, Fontaine V, Kitajima S, Berkhout B. (1999) Evolution of the human immunodeficiency virus type 1 long terminal repeat promoter by conversion of an NF-kappaB enhancer element into a GABP binding site. *J Virol* 73: 1331-40.

Villena, J. A. Martin, I. Vinas, O. Cormand, B. Iglesias, R. Mampel, T. Giralt, M. Villarroya, F. (1994). ETS transcription factors regulate the expression of the gene for the human mitochondrial ATP synthase beta-subunit. J Biol Chem 269(51): 32649-54.

References

Villena JA, Vinas O, Mampel T, Iglesias R, Giralt M, Villarroya F. (1998) Regulation of mitochondrial biogenesis in brown adipose tissue: nuclear respiratory factor-2/GA-binding protein is responsible for the transcriptional regulation of the gene for the mitochondrial ATP synthase beta subunit. *Biochem J* 331 (Pt 1): 121-7.

Virbasius JV, Scarpulla RC. (1994) Activation of the human mitochondrial transcription factor A gene by nuclear respiratory factors: a potential regulatory link between nuclear and mitochondrial gene expression in organelle biogenesis. *Proc Natl Acad Sci U S A* 91: 1309-13.

Vo N, Goodman RH. (2001) CREB-binding protein and p300 in transcriptional regulation. *J Biol Chem* 276: 13505-8.

Vogel JL, Kristie TM. (2000) The novel coactivator C1 (HCF) coordinates multiprotein enhancer formation and mediates transcription activation by GABP. *EMBO J* 19: 683-90.

Wasylyk B, Hagman J, Gutierrez-Hartmann A. (1998) Ets transcription factors: nuclear effectors of the Ras-MAP-kinase signaling pathway. *Trends Biochem Sci* 23: 213-6.

Wasylyk C, Kerckaert JP, Wasylyk B. (1992) A novel modulator domain of Ets transcription factors. *Genes Dev* 6: 965-74.

Watanabe H, Imai T, Sharp PA, Handa H. (1988) Identification of two transcription factors that bind to specific elements in the promoter of the adenovirus early-region 4. *Mol Cell Biol.* 8: 1290-300.

Watanabe H, Sawada J, Yano K, Yamaguchi K, Goto M, Handa H. (1993) cDNA cloning of transcription factor E4TF1 subunits with Ets and notch motifs. *Mol Cell Biol* 13: 1385-91.

Watanabe H, Wada T, Handa H. (1990) Transcription factor E4TF1 contains two subunits with different functions. *EMBO J* 9: 841-7.

Wilson AC, LaMarco K, Peterson MG, Herr W. (1993) The VP16 accessory protein HCF is a family of polypeptides processed from a large precursor protein. *Cell* 74: 115-25.

Wong-Riley M, Guo A, Bachman NJ, Lomax MI. (2000) Human COX6A1 gene: promoter analysis, cDNA isolation and expression in the monkey brain. *Gene* 247: 63-75.

Xue HH, Bollenbacher J, Rovella V, Tripuraneni R, Du YB, Liu CY, Williams A, McCoy JP, Leonard WJ. (2004) GA binding protein regulates interleukin 7 receptor alpha-chain gene expression in T cells. *Nat Immunol* 5: 1036-44.

Xue HH, Bollenbacher-Reilley J, Wu Z, Spolski R, Jing X, Zhang YC, McCoy JP, Leonard WJ. (2007) The transcription factor GABP is a critical regulator of B lymphocyte development. *Immunity* 26:421-31.

Yaneva M, Kippenberger S, Wang N, Su Q, McGarvey M, Nazarian A, Lacomis L, Erdjument-Bromage H, Tempst P. (2006) PU.1 and a TTTAAA element in the myeloid defensin-1 promoter create an operational TATA box that can impose cell specificity onto TFIID function. *J Immunol* 176: 6906-17.

Yang SJ, Liang HL, Ning G, Wong-Riley MT. (2004) Ultrastructural study of depolarization-induced translocation of NRF-2 transcription factor in cultured rat visual cortical neurons. *Eur J Neurosci* 19: 1153-62.

Yang ZF, Mott S, Rosmarin AG. (2007) The Ets transcription factor GABP is required for cell-cycle progression. *Nat Cell Biol* 9: 339-46.

Yoganathan T, Bhat NK, Sells BH. (1992) A positive regulator of the ribosomal protein gene, beta factor, belongs to the ETS oncoprotein family. *Biochem J* 287 (Pt 2): 349-53.

Yokomori N, Tawata M, Saito T, Shimura H, Onaya T. (1998) Regulation of the rat thyrotropin receptor gene by the methylation-sensitive transcription factor GA-binding protein. *Mol Endocrinol* 12:1241-9.

Yordy JS and Muise-Helmericks RC. (2000). Signal transduction and the Ets family of transcription factors. *Oncogene* 19(55): 6503-13.

Yu M, Yang XY, Schmidt T, Chinenov Y, Wang R, Martin ME (1997). GA-binding protein-dependent transcription initiator elements. Effect of helical spacing between polyomavirus enhancer a factor 3(PEA3)/Ets-binding sites on initiator activity. J Biol Chem 272(46): 29060-7.

Zhang C, Wong-Riley MT. (2000) Depolarizing stimulation upregulates GA-binding protein in neurons: a transcription factor involved in the bigenomic expression of cytochrome oxidase subunits. *Eur J Neurosci* 12: 1013-23.

Zhang L, Eddy A, Teng, YT, Fritzler M, Kluppel M, Melet F, Bernstein A. (1995). An immunological renal disease in transgenic mice that overexpress Fli-1, a member of the ets family of transcription factor genes. *Mol Cell Biol* 15(12): 6961-70.

7. APPENDIX

7.1. Abbreviation Index

AEBSF	4-(2-Aminoethyl) Benzenesulfonyl Fluoride Hydrochloride
bHLH-ZIP	Basic-helix-loop-helix Leucine Zipper
BRCA	Breast Cancer
CDKI	Cyclin-Dependent Kinase Inhibitor
CF	Coupling Factor
CH	Colony Hybridisation
cpm	Counts Per Minute
DSE	Dyad Symmetry Element
HCF	Host Cell Factor
HN	Histidine
HRP	Horseradish peroxidase
PMSF	Phenyl Methyl Sulphonyl Fluoride
PNK	Polynucleotide Kinase
Poly dI/dC	Poly (deoxyinosinic-deoxycytidylic) acid sodium salt
TSC	Tuberous Sclerosis
VEGF	Vascular Endothelial Growth Factor

Definitions of Used Terms, Concerning the Growth Properties Of a Cell Culture

In the present work are utilized many terms, dealing with various aspects of growing properties of an (adherent) cell culture. To clarify their meaning used in the text above, a short description is applied.

Postulations of growth/expansion, proliferation and apoptosis:

Cell culture Growth – an increase in the size of the culture by increasing the cell number. This process is the resultant from the sum of processes increasing (Proliferation/Expansion), detaining (Senescence) and decreasing the cell number (Apoptosis/Necrosis). The processes of Senescence and Necrosis are not considered in this work.

Cell/culture Proliferation – the production of cells by multiplication of their numbers.

Expansion of cell culture – the act of spreading out; the condition of being expanded; enlargement of the culture (Typically used when monoclonal cell cultures were created via propagation of an isolated single cell)

7.2. Design of RNAi Targets

Selection of siRNA Target Sites

The most suitable GABP• mRNA target sequences were determined with the help of Ambion's internet tool "siRNA Target Finder" at internet address: http://www.ambion.com/techlib/misc/siRNA_finder.html and according to the following recommendations:

- The target should be selected within 50-150 nt downstream of the start codon.
- 5`and 3`-untranslated regions should be avoided.
- The gene-specific sequence should not contain a stretch of 4 or more A`s or T`s (will give premature termination of the transcript).
- 50% G/C content is optimal (30-70%), highly G-rich stretches to be avoided.
- Intronic sequences should not be included (cellular compartmentalization).
- Preferably, the 23 nt target sequence should be of the following structure: 5`AA-19-TT3`, or at least 5`AA-19-3` (in the last case TT at the 3` end has to be added.
- The selected siRNA target sequence has to be Blast-searched against EST and cDNA (mRNA) libraries.

siRNA target sequences and their positions on GABP• gene sequence

seq.1 (AAGAGAGAAGCAGAAGAGCTG), pos.7; seq.2 (AAGCAGAAGAGCTGATAGAAA), pos.14;
seq.3 (AAGAGCTGATAGAAATTGAGA), pos.20; seq.4 (AAATTGAGATCGACGGGACTG), pos.32;
seq.5 (AAAGCAGAGTGCACAGAAGAA), pos.55; seq.6 (AAGAAAGCATTGTGGAACAAA), pos.71;
seq.7 (AAAGCATTGTGGAACAAACCT), pos.74; seq.8 (AACAAACCTATACCCCAGCTG), pos.86;
seq.9 (AAACCTATACCCCAGCTGAAT), pos.89; seq.10 (AATGTGTAAGCCAGGCCATAG), pos.107;
seq.11 (AAGCCAGGCCATAGACATCAA), pos.114; seq.12 (AATGAACCAATAGGCAATTTA), pos.133;
seq.13 (AACCAATAGGCAATTTAAAGA), pos.137; seq.14 (AATAGGCAATTTAAAGAAACT), pos.141;
seq.15 (AATTTAAAGAAACTACTAGAA), pos.148; seq.16 (AAAGAAACTACTAGAACCAAG), pos.153;
seq.17 (AAACTACTAGAACCAAGACTG), pos.157; seq.18 (AACCAAGACTGCAGTGTTCTT), pos.167;
seq.19 (AAGACTGCAGTGTTCTTTGGA), pos.171; seq.20 (AAATTTGCCTGCAAGATATTC), pos.200;
seq.21 (AAGATATTCAGCTGGATCCAG), pos.212; seq.22 (AAGCTTGTTTGATCAAGGAGT), pos.237;
seq.23 (AAGGAGTGAAAACAGATGGGA), pos.251; seq.24 (AAAACAGATGGGACTGTACAG), pos.259;
seq.25 (AACAGATGGGACTGTACAGCT), pos.261; seq.26 (AATTTCTTACCAAGGAATGGA), pos.294;
seq.27 (AAGGAATGGAGCCAAAGTTGA), pos.305; seq.28 (AATGGAGCCAAAGTTGAACAT), pos.309;
seq.29 (AAAGTTGAACATTCTTGAAAT), pos.318; seq.30 (AACATTCTTGAAATTGTTAAG), pos.325;
seq.31 (AAATTGTTAAGACTGCGGAAA), pos.335; seq.32 (AAGACTGCGGAAACGGTCAG), pos.343;
seq.33 (AAACGGTCGAGGTGGTCATCG), pos.353; seq.34 (AAGCAGAAGCGCATCTCGTTG), pos.395;
seq.35 (AAGCGCATCTCGTTGAAGAAG), pos.401; seq.36 (AAGAAGCTCAAGTGATAACTC), pos.416;
seq.37 (AAGCTCAAGTGATAACTCTTG), pos.419; seq.38 (AAGTGATAACTCTTGACGGCA), pos.425;

Appendix/siRNA Design

seq.39 (AACTCTTGACGGCACCAAGCA), pos.432; seq.40 (AAGCACATTACGACCATTTCA), pos.448;
seq.41 (AAGGCTACAGAAAAGAGCAGG), pos.512; seq.42 (AAAAGAGCAGGAGCGCCTTGG), pos.522;
seq.43 (AAGAGCAGGAGCGCCTTGGCA), pos.524; seq.44 (AAGTCCTGCATTGGGTGGTTT), pos.578;
seq.45 (AATGAAGGAGTTCAGCATGAC), pos.603; seq.46 (AAGGAGTTCAGCATGACTGAT), pos.607;
seq.47 (AACATTTCGGGAAGAGAATTA), pos.646; seq.48 (AAGAGAATTATGTAGTCTCAA), pos.657;
seq.49 (AATTATGTAGTCTCAACCAAG), pos.662; seq.50 (AACCAAGAAGATTTTTTTCAG), pos.676;
seq.51 (AAGAAGATTTTTTTCAGCGGG), pos.680; seq.52 (AAGATTTTTTTCAGCGGGTCC), pos.683;
seq.53 (AAATTCTTTGGAGTCATCTGG), pos.713; seq.54 (AAAATATGTTTTGGCCAGCCA), pos.744;
seq.55 (AATATGTTTTGGCCAGCCAAG), pos.746; seq.56 (AAGAGCAACAGATGAATGAGA), pos.764;
seq.57 (AACAGATGAATGAGATAGTTA), pos.770; seq.58 (AATGAGATAGTTACCATTGAC), pos.778;
seq.59 (AAAGTTATAAACAGCAGTGCA), pos.853; seq.60 (AAACAGCAGTGCAAAAGCAGC), pos.861;
seq.61 (AAAAGCAGCTAAAGTGCAACG), pos.873; seq.62 (AAGCAGCTAAAGTGCAACGGT), pos.875;
seq.63 (AAAGTGCAACGGTCCCCAAGG), pos.883; seq.64 (AACGGTCCCCAAGGATTTCAG), pos.890;
seq.65 (AAGGATTTCAGGAGAAGACAG), pos.900; seq.66 (AAGACAGAAGTTCACCGGGGA), pos.914;
seq.67 (AAGTTCACCGGGGAACAGAAC), pos.921; seq.68 (AACAGAACAGGAAACAATGGT), pos.934;
seq.69 (AACAGGAAACAATGGTCAGAT), pos.939; seq.70 (AAACAATGGTCAGATCCAACT), pos.945;
seq.71 (AATGGTCAGATCCAACTATGG), pos.949; seq.72 (AACTATGGCAGTTTTGCTAG), pos.962;
seq.73 (AACTTCTTACTGACAAGGATG), pos.983; seq.74 (AAGGATGCTCGAGACTGTATT), pos.997;
seq.75 (AAGGTGAATTTAAGCTAAATC), pos.1034; seq.76 (AATTTAAGCTAAATCAGCCTG), pos.1040;
seq.77 (AAGCTAAATCAGCCTGAGTTG), pos.1045; seq.78 (AAATCAGCCTGAGTTGGTTGC), pos.1050;
seq.79 (AAAAATGGGGACAACGTAAGA), pos.1073; seq.80 (AAATGGGGACAACGTAAGAAC), pos.1075;
seq.81 (AACGTAAGAACAAGCCTACCA), pos.1085; seq.82 (AAGAACAAGCCTACCATGAAC), pos.1090;
seq.83 (AACAAGCCTACCATGAACTAT), pos.1093; seq.84 (AAGCCTACCATGAACTATGAG), pos.1096;
seq.85 (AACTATGAGAAACTTAGCCGT), pos.1108; seq.86 (AAACTTAGCCGTGCATTACGG), pos.1117;
seq.87 (AAAGTTCAAGGCAAGAGATTT), pos.1165; seq.88 (AAGGCAAGAGATTTGTGTACA), pos.1172;
seq.89 (AAGAGATTTGTGTACAAATTT), pos.1177; seq.90 (AAATTTGTTTGTGACTTGAAG), pos.1192;
seq.91 (AAGACTCTTATTGGATACAGT), pos.1210; seq.92 (AACTGAACCGTCTGGTCATAG), pos.1238;
seq.93 (AACCGTCTGGTCATAGAGTGT), pos.1243; seq.94 (AACAGAAGAAACTGGCACGGA), pos.1265;
seq.95 (AAGAAACTGGCACGGATGCAG), pos.1270; seq.96 (AAACTGGCACGGATGCAGCTG), pos.1273

Appendix/Clone Charts

7.3. Clone Charts

psiGA#10

(Transcribes siRNA against GABP• mRNA)

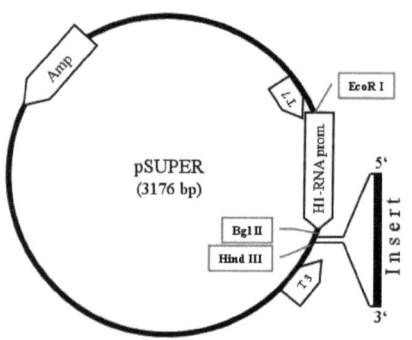

Bgl II and Hind III sites are used for cloning of inserts. Upon ligation Bgl II site is destroyed to give an option to better select positive clones after the ligation.

Position of the insert in gene sequence: 107

Expected transcript:

```
                                    UUC
      5'- UGUGUAAGCCAGGCCAUAG       A
          : : : : :: :: : : : ::::   A
      3'- UUACACAUUCGGUCCGGUAUC      G
                                    AGA
```

Reference: Thijn R. Brummelkamp, Rene Bernards, Reuven Agami. A System for Stable Expression of Short Interfering RNAs in Mammalian Cells, Science, Vol. 296, 19.04.2002

psiGA#7

This plasmid is constructed exactly the same way as psiGA#10, using the corresponding insert.

Insert: 5`-gatccccAGCATTGTGGAACAAACCTttcaagagaAGGTTTGTTCCACAATGCTtttttggaaa-3`

Position in gene sequence: 74

pRetroSuper-GABP• 10

(pRS10)

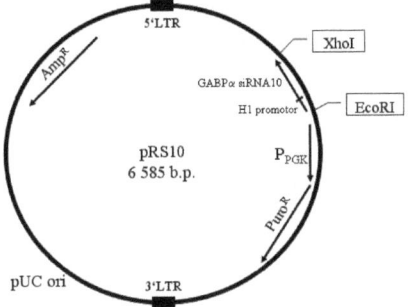

pRS-GABP• 10 is based on **pMSCVpuro** (6,3 kb). XhoI and EcoRI sites are used for cloning of the insert into the MCS (in opposite orientation). Construct is checked with XhoI - EcoRI digestion for presence of the insert. Positive clones should give bands: 6,3 kb, 285 bp. **Insert** derived from pGAsiR10, based on pSUPER (see the clone chart above), digested with the same enzymes and consist of H1 promoter (222 bp) and siRNA sequence #10 (from the targets choosing list) against GABP•

Reference: Brummelkamp et al, Cancer Cell, September 2002, Vol.2, p.243

pMSCVPuro-GFP

(pMSCVpuroG)

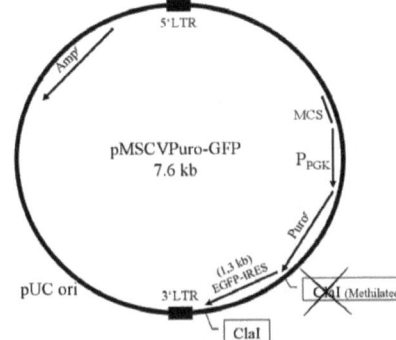

pMSCVPuro-GFP is based on **pMSCVpuro** (6,3 kb). The unique Cla I site is used for cloning of the IRES-EGFP insert. The insert is checked with:

- Cla I digestion. Positive clones should give a linear band.

- Hind III digestion. Positive clones should give bands: 6,7 kb and 892 bp

Insert obtained with PCR on pIRES-EGFP, using primers introducing Cla I sites at the ends.

Primers:

IRES-GFP F2-ClaI: 5'-TAT AT**A TCG AT**C GCC CCT CTC CCT CCC C-3'
IRES-GFP R-ClaI: 5'-ATA T**AT CGA T**GC TTT ACT TGT ACA GCT CGT C-3'

Appendix/Clone Charts

pRetroSuper-GABP• 10-GFP
(pRS10G)

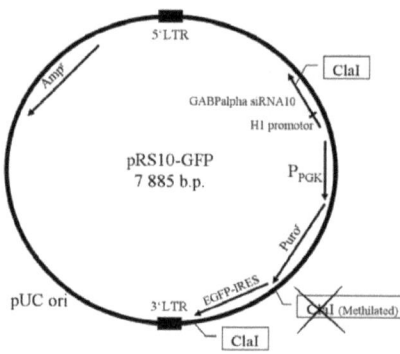

pRS10-GFP is based on pRS10 (6 585 kb). The unique Cla I site is used for cloning of the IRES-EGFP insert. The insert is checked with:
- Cla I digestion. Positive clones should give bands: 5,1kb and 2,8 kb.
- Hind III digestion. Positive clones should give bands: 6,2 kb, 884 bp and 782 bp

Insert obtained with PCR on pIRES-EGFP, using primers introducing Cla I sites at the ends.

Primers:
IRES-GFP F2-ClaI: 5'-TAT ATA TCG ATC GCC CCT CTC CCT CCC C-3'
IRES-GFP R-ClaI: 5'-ATA TAT CGA TGC TTT ACT TGT ACA GCT CGT C-3'

pMSCVpuro-delta
(pMSCVp-•)

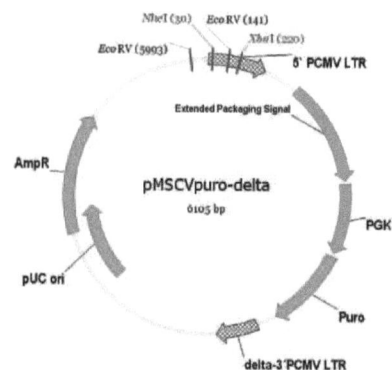

pMSCVp-• is based on pMSCVpuro (6.3 kb). The plasmid is constructed by full digestion with Sac II, followed with digestion with Sca I. The smaller band, containing the 3'LTR was cut with Nhe I and Xba I. The two bigger bands from this digestion were isolated and triple ligation was performed together with the bigger band from the Sac II/Sca I digestion. Construct was checked with Eco RV digestion. Positive clones give bands: 5,8 kb, 253 bp

Reference: Brummelkamp et all, Cancer Cell, September 2002, Vol.2, p.243

pTRE-GABP•

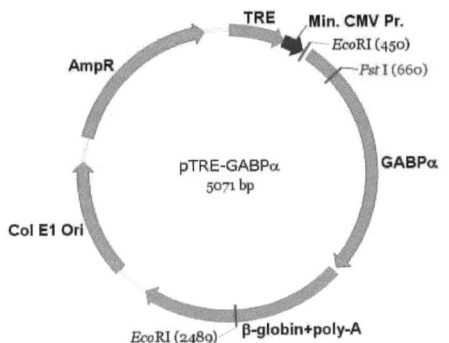

The plasmid is constructed on the basis of pTRE-6xHN, cut with BamHI + XbaI. The insert is derived from PCR fragment, amplified with primers: mGAa5'BglIIpure (F) and mGAa-3' XbaI (R) from plasmid pEGZgabpa-6 and cut with the same restriction enzymes.

pTRE-GABP•

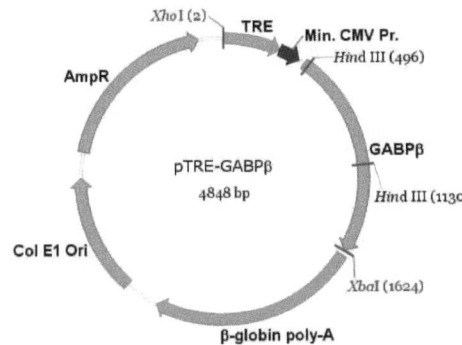

The plasmid is constructed on the basis of pTRE-6xHN, cut with BamHI + EcoRV.
The insert is derived from PCR fragment, amplified with primers: mGab1-5'-BglIIpure (F) and GABPb-HpaI (R) from pRSV-GABP•$_1$ and cut with restriction enzymes HpaI + BglII.

pMSCVpuroG-GABP•

(Retroviral vector, constitutively expressing GABP• and EGFP)

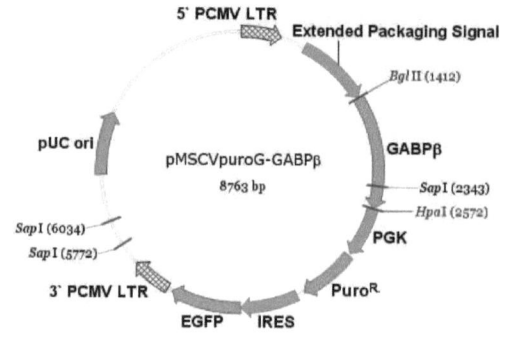

The plasmid is constructed on the basis of pMSCVpuro-EGFP (pMSCVpuroG), cut with BglII + HpaI. The insert is derived from PCR fragment, amplified with primers: mGab1-5'-BglIIpure (F) and GABPb-HpaI (R) from pRSV-GABP•₁ and cut with restriction enzymes HpaI + BglII.

pMSCVpuro-•-TRE-GABP•

(Self-inactivating retroviral vector, Expressing GABP• under the regulation of TRE)

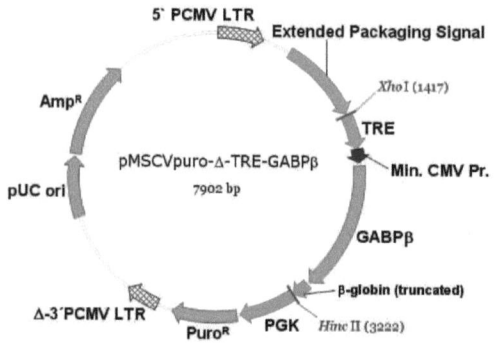

The plasmid is constructed on the basis of pMSCVpuro-•, cut with XhoI + HincII. The insert is derived from pTRE-GABP•, cut with the same restriction enzymes.

pMSCVpuro-•-TRE-GABP• ••

(Self-inactivating retroviral vector,
Expressing both GABP• and GABP• under the regulation of TRE)

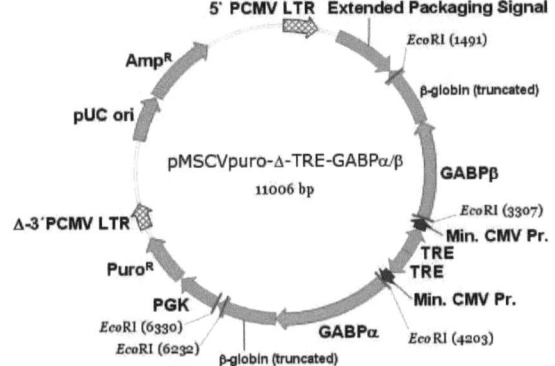

The plasmid is constructed on the basis of pMSCVpuro-•, cut with BglII. The insert is derived from cutting and ligating of TRE-GABP• and -• cassettes from the plasmids TRE-GABP• and TRE-GABP• with BglII + XhoI.

i want morebooks!

Buy your books fast and straightforward online - at one of world's fastest growing online book stores! Environmentally sound due to Print-on-Demand technologies.

Buy your books online at
www.get-morebooks.com

Kaufen Sie Ihre Bücher schnell und unkompliziert online – auf einer der am schnellsten wachsenden Buchhandelsplattformen weltweit! Dank Print-On-Demand umwelt- und ressourcenschonend produziert.

Bücher schneller online kaufen
www.morebooks.de

VDM Verlagsservicegesellschaft mbH
Heinrich-Böcking-Str. 6-8
D - 66121 Saarbrücken

Telefon: +49 681 3720 174
Telefax: +49 681 3720 1749

info@vdm-vsg.de
www.vdm-vsg.de

Printed by Books on Demand GmbH, Norderstedt / Germany